Building an Industry

A History of Cable Television and its Development in Canada

K.J. Easton

Pottersfield Press
Lawrencetown Beach
Nova Scotia, Canada

Copyright © 2000 Kenneth J. Easton

All rights reserved. No part of this publication may be reproduced or transmitted in any form or by any means, electronic or mechanical, including photocopying, or by any information storage or retrieval system, without permission in writing from the publisher.

Canadian Cataloguing in Publication Data

Easton, K.J. (Kenneth J.), 1916–

Building an industry

ISBN 1-895900-28-X

1. Cable television — Canada — History. I. Title

HE8700.72.C3E37 2000 384.55'5'0971 C99-950245-X

Cover illustration: Gail LeBlanc

Pottersfield Press gratefully acknowledges the ongoing support of the Nova Scotia Department of Tourism and Culture, Cultural Affairs Division, as well as The Canada Council for the Arts. We acknowledge the financial support of the Government of Canada through the Book Publishing Industry Development Program for our publishing activities.

Printed in Canada

Pottersfield Press
Lawrencetown Beach
83 Leslie Road
East Lawrencetown
Nova Scotia Canada B3Z 1P8
To order, telephone:
1-800-NIMBUS9 (1-800-646-2879)

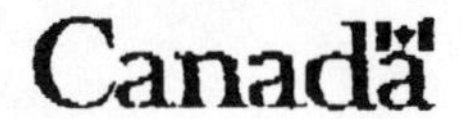

Contents

INTRODUCTION

In 1980 Ken Easton wrote and published *Thirty Years in Cable TV: Reminiscences of a Pioneer*. The book was inspired by the formation in 1976 of a Canadian Cable Television Association Pioneers' Club and the realization of the need to record some of the history of the birth of this industry while those who participated were still around to recount it. Planning for the future must be guided by what has happened in the past if we are to avoid repeating past mistakes; it is only by looking back over the past history that we are able to realize what tremendous changes have taken place in a relatively short time, and thus provide a measure of the further changes which can be expected in the future.

While partly autobiographical, since the author had been directly involved in this history for more than fifty years, the book was intended to record the early history of cable TV in North America, and particularly in Canada, from the point of view of some of the early pioneers, who, like him, had been personally involved in its creation. At that time cable television, being used in less than 25 percent of the homes in North America, was not as familiar to the general public as it is today, and it was thought that the main interest in this history would be among those directly and indirectly engaged in cable TV and broadcasting. Accordingly only a limited edition of the book was published and it was not available to the general public through the retail book trade.

Now, twenty years later, the industry has grown to the point where a majority of the TV homes in Canada and the United States receive greatly expanded television service by cable. Cable TV is now almost as ubiquitous in the North American home as the telephone, and program services designed specifically for cable distribution, such as Cable Network News (CNN) and CBC Newsworld, are as well known as the daily newspaper.

Therefore, the time has come for a book which can tell this story to the general public, and at the same time bring it up to date by covering the major changes which have taken place since 1980, particularly in the technology. These include the introduction of pay-TV, the use of satellites for television distribution nationally and internationally with the resulting proliferation in channels and cable-specific services, and more recently the beginnings of high definition television, digital video compression, and the inexorable coming together of cable TV and telephone technologies. This book also describes the origins of the Canadian Cable Television Association, its growth with the industry it serves, and its influence as a medium between the industry and its regulators through the years.

The first chapters reiterate much of the ground covered in the previous book up to the early 1970s, but it is rewritten as a history rather than an autobiographical reminiscence. The story is then continued up to the present day, and a complete chapter has been added describing the evolution of the technology. The earlier book exercised a rigorous avoidance of technical terminology or references to the technical aspects in deference to the average reader; however, cable TV is very much a technology-based industry, and throughout most of its development the various stages of this evolution have been technology-led. This history therefore would not be complete without reviewing the evolution of the technology which was the lifeblood of the industry's remarkable growth to maturity in little more than five decades.

CHAPTER 1

Pre-history: radio and television relay in the U.K.

The idea of distributing broadcast programs by cable to individual homes in a community as an alternative to direct reception at each home was not new even in 1948 when the earliest cable systems for television distribution were being developed in North America. By that time nearly 900,000 homes in the United Kingdom received radio programs by relay, the term given to cable systems distributing these programs from a central receiving location. These homes represented some 7.5 percent of all the licenced radio receivers in the U.K. and relay systems had been in operation on a commercial scale since the mid-1920s.

In addition to this extensive development of radio relay in the U.K., by the late 1940s there were also cable distribution systems of significant size in several countries on the continent of Europe, notably Belgium, the Netherlands, Russia, Switzerland, Sweden, and Germany. In most cases the relay systems used a separate copper-pair cable network distributing the program material at audio frequencies, but in Sweden, Germany and parts of Switzerland, they used high-frequency carriers on the regular telephone lines. There were also relay systems in several of the British colonies of that time: Nigeria, the Gold Coast (now Ghana), and Malta.

In fact in Malta there was no broadcasting as such. The radio programs were distributed throughout the island entirely by wire. This was probably just as well considering the intensive and prolonged air attacks this island fortress suffered during World War Two before the occupation of Italy, which earned the whole community the George Cross — the most coveted and meritorious British civilian decoration for bravery under fire. Also around this time there was further development of radio relay in British overseas territories, particularly in Hong Kong where construction started on a system which became the most extensive of its kind in the world and later included television distribution.

The reason for this early development of wire distribution when broadcast coverage, at least of most major residential areas, was adequate was a matter of cost and convenience to the listener. The majority of the radio receivers in the early days of broadcasting were crystal sets. These used a crystal in its natural state consisting of a sulphide of lead, copper, or iron. The fine point of a wire, known as a cat's whisker, was brought into contact with the crystal, and when a sensitive spot on the surface had been found this would cause detection of the radio signal. It is interesting to look back from these days of transistor radios and solid-state electronics in all its ramifications, and realize with hindsight that this was probably the first practical application of solid-state physics, and that the crystal detector was in fact a semi-conductor point-contact diode.

These receivers simply detected the radio signal and had no means of amplifying either the radio signal or the detected audio obtained from it by the crystal. In consequence their sensitivity was very low, providing only sufficient power to operate a pair of earphones, and then only if the listener was close enough to a broadcast transmitter to pick up a sufficiently strong signal.

By the mid-1920s receivers using vacuum tubes to provide both signal detection and amplification had been developed and were coming into fairly general use. These receivers were both cumbersome and expensive since they were not capable of being operated from the domestic power supply, and it was necessary to power them from batteries. No less than three batteries were required for this purpose: a low-tension lead-acid accumulator to heat the filaments; a high-tension battery for the plates or anodes; and a grid bias battery to adjust the operating condition of the tubes. The low-tension battery was rechargeable, but the other two each comprised an assembly of small dry cells in a block connected in series since the voltage required was higher but the current drain was much lower than that required for the filaments.

Apart from the higher cost of these vacuum tube receivers, the additional cost of the batteries and the need for regular recharging or replacement made these sets considerably more expensive and inconvenient for the user. However, the use of tubes provided considerable amplification of the signal so that the receiver sensitivity was greatly improved, and it was possible to generate enough audio power to be able to drive a loudspeaker instead of having to rely on earphones. The early loudspeakers took the form of a vibrating diaphragm, very much like a glorified earphone, coupled to a horn or trumpet similar to those used on the early phonographs. A major advantage of these early battery-operated receivers with loudspeakers, of course, was that it was possible for the broadcast program to be heard by several people in the room at the same time instead of having to take turns with the earphones!

It wasn't long before someone figured that it should be possible for several people to hear the programs simultaneously in different rooms, or even in different homes, if the radio receiver could be used to drive several loudspeakers and if the speakers could be located away from the receiver and connected to it by wire. One of the earliest recorded instances of this involved Wallace Maton in the small town of Hythe, near Southampton, in southern England. Maton, who owned an electrical shop in Hythe and also ran the local cinema, was greatly interested in radio and had built a battery-operated receiving set. So his wife could hear the programs when she was in another part of the house, he connected the set by a pair of wires to a loudspeaker in another room as an experiment. Finding that this was successful, he decided to investigate the possibility of extending the wire for longer distances.

In fields at the back of his house he ran out a length of wire to a distance of half a mile and, connecting a loudspeaker to the end, found that the broadcast programs were reproduced with little if any loss of volume. He then found that this was also the case if several loudspeakers were connected to the wire, and these results encouraged him to carry his experiments further. He arranged with friends in the town to allow him to install loudspeakers in their homes which he then connected by wire to the receiving set in his own home. These friends were then able to hear the broadcasts without possessing a receiving set themselves.

As no insurmountable difficulties had been encountered Maton decided that it should be possible to develop this system of receiving broadcast programs and distributing them by wire on a commercial basis. He therefore began to charge one shilling and sixpence per week (approximately nineteen cents) for his service and extended his wire system to serve additional subscribers. In this way the first commercial relay system in Great Britain was started in January 1925. This system continued in operation until 1941 when Maton closed it down because of shortages of material and labour during the war. There were never more than 150 subscribers, but the system was remarkable not only because it was the first, but also because it transmitted the audio over considerable distances and covered an area with a very low population density. In fact the subscriber furthest from the receiver required ten miles of wire to reach him.

Maton's system in Hythe was the first significant example in the United Kingdom of the commercial distribution of broadcast programs by wire because it led to the official recognition of relay services and set an example of what could be done, which was to be widely followed by others. However, it was not the first recorded example of radio relay, and indeed was not the first such system to be operated on a commercial basis.

It is probable that the first commercial system preceded Maton's by some nine months in the Netherlands. Andrianus Bauling was experimenting with radio reception in the very early 1920s in Koog aan de Zaan, a small town northwest of Amsterdam. Friends and neighbours were so intrigued by the results he was getting that the Bauling home became a neighbourhood centre requiring large quantities of free coffee in conformity with traditional Dutch hospitality. He therefore ran a wire to the home of his most pressing neighbour and later extended it to others in the vicinity, charging each the equivalent of three shillings, or approximately thirty-eight cents, per month. Each subscriber listened to the programs using headphones, some having a pair for each member of the family, and they were notified that there was something to listen to by the ringing of a bell in their homes.

This first commercial radio relay service in the Netherlands, and possibly in Europe, started in Koog aan de Zaan on April 16, 1924, with a broadcast of the St. Matthew Passion from the St. Bavo church in Haarlem. Later Bauling built other commercially successful relay systems in the Netherlands, and in 1931 and 1932 he built systems in Nottingham and in Wolverhampton, England, which were sold to Rediffusion in 1935. The Dutch systems were forcibly nationalized in 1940 during the German occupation, and by that time more than a quarter of the homes licenced for radio reception were relay subscribers.

It is clear that by 1925 the feasibility of relaying broadcast radio programs by wire to individual homes in a community from a central receiver had been well established, and the circumstances of the day made it an attractive proposition for potential subscribers. Radio as a news and entertainment medium was very young and, like most innovative services of its kind, had an attraction for the public. The fact that one could listen to plays, concerts, news reports, etc. while they were actually taking place many miles away held a fascination and provided an element of immediacy which improved on the phonograph and the newspaper.

However, until the introduction of relay it was necessary to buy a receiver in order to use this service. Since these were either inefficient crystal sets giving generally poor reception and limited to one listener with a pair of headphones, or vacuum tube sets which were expensive and required regular recharging or replacement of batteries, a relay connection at one shilling and sixpence per week with no capital outlay was relatively attractive. Furthermore, this charge was less than the weekly instalment payments on even the cheapest radio receiver available at that time. Consequently, relay services developed and flourished in many communities in the U.K., and one of the companies which became deeply involved in this development was Rediffusion Ltd. This company was formed in the late 1920s by four radio pioneers who had the manufacturing and marketing rights to a loudspeaker which was particularly suitable for relay use. By 1946 there were nearly three-quarters of a million homes in the U.K. receiving their radio programs by relay, and of these Rediffusion had more than one-third connected to systems operating in thirty-one separate communities in England and Wales.

In spite of material and labour shortages during World War Two, from 1939 to 1945, and notwithstanding Maton's difficulties in Hythe, relay continued to grow. Radio was an essential service to the public in wartime England, not only as a source of news but as a source of entertainment and diversion from the worries of air raids, rationing and, in the first two or three years, serious reverses and disasters — in fact a badly needed morale booster. As a broadcast medium, its reliability could not be taken for granted because raiding enemy bombers could use the radio transmitters as homing beacons to guide them to their targets, and it was frequently necessary to have them shut down or reduce power at very short notice for hours on end, and frequently night after night. Under these circumstances relay was much more reliable, especially as many of the larger systems by this time did not rely on off-air reception of the broadcast programs but had direct line connections to the nearest studio or program centre.

A brief description of a radio relay system as it had developed by the mid-1940s might be in order here, particularly since it will be pertinent to a later discussion of the early development of television relay.

First it must be understood that the radio programs were not distributed on the cable at the radio frequencies on which they were received, as is commonly the case with television distribution by cable. Since the original motive for relay was to simplify the subscriber's equipment by removing every part of a receiver except the loudspeaker, the programs were distributed at audio frequencies from a central receiver. Thus, all a subscriber needed for reception was a loudspeaker and a volume control. Most relay systems distributed at least two or three programs, each program being carried on a separate pair of wires in the cable so that the subscriber's installation would also include a simple rotary switch for program selection between the pairs. Typically the loudspeaker would be of nine-inch diameter, with a high-impedance input, mounted in a plastic cabinet of pleasing appearance. Since all the audio power necessary to drive the speaker was obtained from the relay cable, no electronic equipment was required in the subscriber's home, and hence there was no need for connection to a domestic power supply.

This was another reason for the popularity and growth of radio relay in the U.K. at that time. In the 1920s and '30s many homes in the industrial, manufacturing, and commercial centres had no domestic electricity supply. The homes were generally heated by coal fires, and town gas made from coal was used for lighting and cooking. In these areas, which were mostly low-income areas, even if the residents could afford the new so-called "battery eliminators" or "all-mains receivers" which did not need batteries, they could not be used because there was no power supply available. For these reasons relay enjoyed its greatest growth in these so-called "working class" areas, and some of the largest systems operated by companies such as Rediffusion were in cities like Newcastle, Sunderland, Hull, Nottingham, and Plymouth, and in South Wales mining and steel centres like Merthyr, Rhondda, Newport, and Swansea.

A typical relay cable of the post-war period consisted of two copper-pairs, not unlike telephone pairs, insulated with polythene, and laid up in "star-quad" formation: that is four wires twisted together forming the points of a four-pointed star, the diagonally opposite wires forming the two pairs. Since each pair could carry one audio program, one quad cable could carry two programs, and if three or four programs were to be distributed then two such cables would be installed side by side. This type of cable developed directly from the invention of polythene during the war and was a considerable improvement on open-wire line and paper- or rubber-insulated twisted pair cable of the earlier period, and soon became the standard cable for radio relay.

The broadcast programs were received off-air and demodulated to audio, or in some cases they were received at audio from the program source over direct lines rented from the telephone authority at that time, the Post Office. The audio signals were then amplified by large power amplifiers and transmitted on the relay system cables at a level of some sixty volts. The amplifiers typically had a power output capacity of the order of a thousand watts, and since a subscriber's installation with the high-impedance loudspeaker did not require more than about one-quarter watt even at maximum volume, one set of amplifiers was sufficient to feed several thousand subscribers.

In the early days the economics were such that it paid to spend a lot of money developing a loudspeaker to obtain maximum sensitivity by using large units, high magnetic flux densities, and large cabinets for acoustic efficiency, and the resulting special relay loudspeaker would then give acceptable volume with the small quarter-watt input. The alternative of using standard loudspeakers with lower sensitivity and more copper in the lines to reduce distribution losses was less economic. This was undoubtedly a major reason for the founders of Rediffusion, who had the rights to just such a high-sensitivity loudspeaker, to get into relay in the first place.

Around 1934 two men who had been involved in relay in Nottingham formed a company to provide a form of relay service to an entirely different market in London. They originally intended to sell expensive radio equipment to the affluent occupants of high-class and expensive apartments in the West End of London, and this intent was expressed in the name of their company, Radio Furniture and Fittings. The equipment was custom-designed to match the interior decor of each apartment, but they soon found that this market was inhibited by the difficulty of obtaining adequate radio reception in the presence of high electrical interference levels without a good antenna, and the corresponding difficulty of installing a good antenna in an apartment. They therefore started to install cable systems in these buildings which could distribute radio programs from a central antenna located on the roof.

In these systems the motive for providing the service was different from that underlying normal radio relay. Rather than replacing much of the subscriber's receiving equipment and leaving only a loudspeaker, with distribution from a central receiver over the cable at audio frequencies, in these systems the broadcast signals were distributed by cable as received from a communal antenna to individual receivers. Such an installation would involve a long-wire antenna, suitable for receiving medium waveband radio signals, erected on the roof of the building, feeding into an amplifier with a bandwidth broad enough to accommodate the entire radio band. The amplifier in turn would feed coaxial cables, generally run in vertical risers on the outside of the building, with taps into the individual apartments for connection to the receivers. This was probably the first application of coaxial cable, another wartime development, to relay, and the precursor of today's master antenna television (MATV) systems, which perform the same function in most multiple dwelling buildings for the provision of broadcast signals to the TV sets installed in the individual units.

By 1939, when World War Two started in Europe, the company had wired some thirty or forty buildings in London serving some 3,000 to 4,000 apartments. This service was continued during the war because of the essential nature of radio service, although it could not be expanded. Shortly after the war ended in 1945 Rediffusion, which was anxious to develop a relay service in London, took over Radio Furniture and Fittings and renamed it London Rediffusion Service Ltd. Since the existing antenna service in the blocks of apartments was the nucleus of this business, this appeared to be the natural place to commence this development.

In the U.K. there is no ready access to utility poles for attachment of relay cables as there is in North America, since in residential areas the majority of the public utilities are placed underground. The usual method of relay installation was by attachment to private property. In a typical residential area which, in urban and suburban areas at least, usually consists of either closely-spaced semi-detached houses or rows of terrace housing, the relay cable was generally run along the faces of the buildings, attached to the brick walls in an unobtrusive position below the eaves. Rights were obtained from the municipal authority to span overhead where possible from building to building for the purpose of crossing streets.

In order to attach to private property in this manner it was necessary to obtain a wayleave from each and every property owner involved, a tedious and time consuming procedure, and these wayleaves were legally validated by payment of a so-called "peppercorn rental" of one shilling (approximately twelve cents) per year. Very little wiring of this type is undertaken now since most current residential construction in areas served by relay comprises multiple dwellings or municipal housing, and the wiring is included with the construction, while cable installation in new franchise areas generally requires underground construction.

However, installing relay cable in the West End of London was in no way comparable to installing in the closely-packed residential areas in which most of Rediffusion's service had been provided in the past. In the first place the apartment blocks in which antenna services were installed were not contiguous but rather randomly located within an area of several square miles. Connecting these buildings together to form any sort of network to be fed from a central source was greatly complicated by the difficulty in crossing streets in this part of London. For example, there were no less than thirteen authorities to be consulted for approval in order to cross a street with a cable, and this would have to be underground at very considerable cost.

Since a relay system, by its very nature, requires each subscriber to be connected to a central program source, and since a cable network over or under the streets was impractical, the problem was solved by renting from the Post Office music-grade telephone lines from the distribution centre to each building or group of buildings to be served, one for each program to be distributed. Each of these lines was terminated in the building to be served on a small power amplifier of twenty to one hundred watts capacity, depending on the number of potential subscribers in the building. Distribution within the building was then accomplished by normal relay cabling, generally run in vertical risers on the outer walls alongside the existing coaxial cable being used for the radio antenna service.

In fact separate lines were not rented from the distribution centre direct to every building, since this would have been uneconomic and would undoubtedly have exceeded the capacity of the Post Office line plant at this location. Instead buildings in the same general area were grouped together and one of them, centrally located within the group, would be designated a repeater station, having a direct connection from the program centre and shorter lines radiating out to the other buildings in the group. Thus a "hub" type of distribution system was established and applied to radio relay thirty years before the same principle was applied to large cable television systems in North America. By mid-1947 nearly one hundred buildings in Central London had been wired with quad cables for relay service, together with coaxial cable for radio antenna distribution. Most of these were large apartment buildings, but they included some of the better-known hotels in the West End of London.

The method of wiring these buildings is of interest since this particular technique is not known to be in use today. In most cases it was either impractical or unacceptable to run the cable inside the building, and it was necessary to install it externally. None of the buildings was high-rise as we understand the term today, but the majority were of at least five to ten storeys and therefore too high for access by ladder. The cables were consequently run in vertical risers on the outside face of the building, usually passing the living-room windows in each apartment for ease of drop access, and were stapled to the brickwork. The plastic cable jackets were generally of a colour to match the colour of the bricks so as to be as unobtrusive as possible.

To avoid the necessity of gaining access to every suite while the feeders were being installed the work was done from bosun's chairs suspended from special temporary rigging on the roof, somewhat in the manner of a window washer on a high-rise building today. The installer would sit in the bosun's chair and be slowly lowered down the line of the riser, tacking the cable in place and installing tap-off units outside every apartment as he went. For this purpose the cable company employed a full-time ex-navy rigger who was kept busy erecting and dismantling the rigging on the buildings being installed, and on whose skill the safety of the installers literally depended.

On November 2, 1936, the British Broadcasting Corporation (BBC) officially opened the first television transmitter for regular public service, located in an old Victorian exhibition building, the Alexandra Palace, in North London. This was the culmination of some thirteen years of development work in England by John Logie Baird.

Baird started his experimental work on television in 1923 and gave the first demonstration of his system in London before an invited audience of members of the Royal Institution on January 26, 1926. Six months later, on August 5, he received a licence to operate a transmitting station on 200 metres (1,500 KHz), with a power of 250 watts and the call sign 2TV. However, the power authorized was inadequate for Baird's purpose and he entered into lengthy negotiations with the BBC and the Post Office, which was the licencing authority, for the part-time experimental use of one of the BBC's radio broadcasting transmitters. This was possible because his system, which used mechanical scanning based on the Nipkow rotating disc, employed only thirty lines and could therefore operate on a relatively low carrier frequency within the medium wavelength radio band.

Baird finally made his first broadcast over 2LO, the BBC's transmitter located in the West End of London, on September 30, 1929. This was an historic occasion which was not matched by the primitive simplicity of the facilities used. Since the technology had not yet developed to the point of being able to combine vision and sound together on one carrier, the transmission first provided two minutes of picture only and then the verbal contribution followed, accompanied by a blank screen! In March 1930 Baird was able to transfer his experimental broadcasts to the BBC station at Brookmans Park, where twin transmitters were available and it was possible to transmit vision on one and sound on the other, thus making a simultaneous sound and vision service possible for the first time.

During this experimental phase Baird's company had been originating its own program material, but in August 1932 the BBC took over the programming. Subsequently the Baird company developed and demonstrated a system based on 240 lines with twenty-five frames per second; however, an EMI development team headed by Schoenberg, Blumlein, and White were also working on a mechanical scanning system, initially using one hundred lines and later extended to 243 lines, and were pressing the BBC to install their equipment for practical tests.

The Cossor company was also developing an alternative system, although it soon dropped out of contention. In the light of the competing claims of these systems the Postmaster General appointed a television advisory committee, known as the Selsdon Committee, to advise the government on the future of the new service. In the meantime, EMI had reported progress on a system using electronic scanning, based on the newly developed Emitron camera tube, which eliminated the rotating disc, making a higher standard of resolution possible, and subsequently offered a 405-line system to the BBC for trial.

The Selsdon Committee was reluctant to make a choice between these two systems and recommended that a public service should commence using the Baird system and the EMI system on alternate days. The public television service was opened on November 2, 1936, on this basis, but after three months of broadcast operation the 405-line electronic system proved to be superior and the Baird system was dropped in February 1937. The EMI system thus became the British standard for at least the next thirty years.

Inevitably, as with any new service to the public requiring newly developed equipment in the hands of the user, the receivers were large and expensive. Thus, among the first purchasers were some of the affluent occupants of the London apartments in which radio antenna systems had been installed. It soon became evident to the company providing this service that similar problems existed in receiving television signals inside an apartment building without a suitable antenna, so they started to add a roof-mounted television antenna and amplifier to each of their coaxial cable radio distribution systems. In this manner cable television was born in 1937, almost as soon as the first public television service in the world started.

By comparison with even small or fairly simple cable television systems of today, these early systems were basic in the extreme. They were essentially master antenna television (MATV) systems, as we now call them, serving only single multiple-dwelling buildings, or at the most groups of adjacent buildings, as distinct from cable TV systems which install cables on public rights-of-way and serve communities of individual homes. Furthermore, only one channel was distributed (channel 1 with a vision carrier of 45 MHz), because that was all that was available and would be for nearly twenty years. Nevertheless, they incorporated all the basic elements of a cable TV system, including the use of coaxial cable as a matched transmission line and broadband high-frequency amplifiers. Although the amplifiers used were suitable for only a single channel, in those days they were broadband amplifiers since a bandwidth of at least 8 MHz at a frequency of 45 MHz was a state of the art achievement for that time.

With war imminent the public television service from Alexandra Palace closed down on September 1, 1939, and did not resume again until June 7, 1945, after the end of the war in Europe. At this time the television antenna service in these apartment installations was reactivated, and in September 1945 their operation was taken over by London Rediffusion Service.

Rediffusion's original interest in television was in the retail market for receivers, as it realized there would be a considerable and growing demand once the public broadcasting service was resumed after the war. As early as 1943, as part of the preparation for post-war resumption of civilian business, the company had been working on the design of television receivers and was ready for limited manufacture by 1947. The first production models used a nine-inch round picture tube, virtually an extension of the type of tube used for oscilloscopes and the early radar displays, but by 1948 a twelve-inch console receiver was also in production.

The first of the nine-inch sets were delivered to London Rediffusion on November 14, 1947, just in time to be tested and set up in the company's offices and showroom for viewing by invited guests of the televised wedding of Her Royal Highness Princess Elizabeth and Prince Philip on November 20. The first of the twelve-inch models were delivered in January 1948.

Although Rediffusion's original intention had been to manufacture television sets for sale this was not really in keeping with the philosophy of relay, which was to simplify the equipment required by a subscriber and supply a service for a regular payment with no capital outlay by the subscriber. Thus, the policy was changed to rental of the receivers, complete with service, to relay subscribers rather than making a one-time sale. It is significant that this marked the beginning of not only a company policy for Rediffusion, but also a trend which has continued to this day in the United Kingdom throughout the whole television retail industry, where receiver rental is generally preferred by the public to outright purchase. In January 1948, with the delivery of the first of the twelve-inch models, London Rediffusion was able to commence offering a receiver rental service to its antenna service subscribers in the buildings where television distribution was available.

In early 1947 London Rediffusion completed a radio relay installation in the London Clinic, a high-class and expensive nursing home on Devonshire Street in the heart of the West End and just around the corner from Harley Street, the well-known address for the top medical specialists. This nursing home has subsequently been patronized by many well-known personalities, including Elizabeth Taylor, and more recently in 1998 by General Augusto Pinochet, the ex-dictator of Chile who was detained there by the British government following medical treatment while awaiting a legal decision on his immunity from prosecution for crimes against humanity.

At that time the London Clinic was a large building consisting of a basement and nine floors. The basement and the first floor contained administrative and domestic services, and the eighth and ninth floors contained operating and treatment rooms. The remaining six floors provided accommodation for patients in a total of 196 private rooms. The whole building had been wired for four-program radio relay service using four-pair cable, and the installation had proved extremely difficult because external wiring was not permitted and all internal wiring had to be hidden and as unobtrusive as possible. Even the standard loudspeakers in plastic cabinets, normally used for relay service, were not allowed because they would not have harmonized with the decor of the rooms, and it was necessary to have special bedside cabinets manufactured which included the loudspeaker and controls within their structure. The controls were easily accessible from the adjacent bed, yet the whole assembly did not obviously contain a loudspeaker and wiring.

Soon after Rediffusion began the receiver rental program, the management of the London Clinic inquired about the feasibility of installing a television distribution system in the building, on a limited scale at first and possibly extending later to all the rooms. The Clinic was interested in the possibility of renting TV sets to patients, and had received requests for TV antenna facilities from patients bringing their own sets; however, it had no idea what sort of demand might be created at the rentals it would be necessary to charge, and without this information the complete rewiring of the building with coaxial cable could not be justified. Rediffusion agreed to provide an antenna service as economically as possible to about 20 percent of the rooms and supply a small number of sets so that rentals could be offered to a limited number of patients for a period and so form an estimate of the likely demand before proceeding with any more extensive installation.

On the face of it, this seemed a fairly simple commitment, and indeed would have been if coaxial cable could have been run externally to temporary outlets in some forty rooms grouped conveniently close together. In spite of the experimental nature of the initial installation, the Clinic management was still not prepared to allow external wiring, particularly of a temporary nature, but if the cable had to be installed internally in a concealed manner like the radio relay cable it was unlikely that it could be done on an experimental basis without excessive cost.

Rediffusion decided to investigate the possibility of using the existing relay cable to distribute the high frequency television signals together with the audio programs, especially as only three programs were being distributed at the time so that the fourth pair was spare and could be used if it were feasible. In order to determine the feasibility of using copper pairs, which were designed and intended for audio transmission, at the frequencies involved in television transmission, around 45 MHz, it was necessary to conduct tests to find what the cable losses were at those frequencies. The resulting information was then applied to the lengths of cable which would be required in the London Clinic to arrive at a network layout which would deliver sufficient signal to each of the television receivers to be connected to the system.

At the London Clinic a pool of receivers was to be available for rental to patients, so that any one receiver might be used on any cable outlet in the building and could be moved from outlet to outlet at fairly frequent intervals. It was, therefore, essential for operational convenience that the variations in signal level between outlets be limited to avoid the necessity of making internal adjustments each time a set was moved. Special tap-off devices, each consisting of a balanced resistive network, were designed to connect the outlet in each room to the distribution cable and deliver the signal level required by the receivers within these limits.

The layout of the existing radio relay system was such that there was a single vertical riser run in a service duct from amplifiers on the first floor to each of the six accommodation floors, with horizontal feeders on each floor serving the individual rooms. For TV purposes it was desirable to have more than one vertical riser in order to avoid the extra losses involved in the network which would be necessary to connect each horizontal feeder. Since there were four pairs in the relay cable, three of which were carrying audio programs, it was decided to experiment with combining the audio and the TV signals together on the same pair. Special filter networks were designed to permit the addition of TV to each pair in the amplifier room so that the TV signal was carried on all four pairs, and similar filter networks were used to select one of the pairs at each floor and divert the TV signal into the horizontal feeders. Tests indicated that there was no mutual interference between the TV and audio signals so that this arrangement was practical.

With the details of a possible system worked out, it was explained to the Clinic management that, although experimental, if successful this could result in substantial cost savings and greatly reduce inconvenience to both patients and staff by avoiding further cabling. It was proposed that the installation be done in three stages: (1) an experimental installation to a limited number of rooms on a single floor, with the provision of a small number of receivers to cooperative patients free of charge; (2) planning of a practical system for the whole building and extension on a more permanent basis to about 20 percent of the rooms so that the Clinic might carry out their trial rental offering; (3) if both the technical and the rental trials were successful then service would be extended to all remaining patients' rooms in the building.

The first stage of the installation was successfully completed by the end of 1949, and it was demonstrated that the ideas which had been developed on a theoretical basis could in fact be applied to an existing radio relay network. The second stage then proceeded and was completed by March 1950 to the entire satisfaction of the London Clinic and of Rediffusion, since the company could foresee considerable possibilities in the system if it could be applied on a larger scale to existing relay systems using balanced-pair cable and so avoid the need to install coaxial cable in order to add TV distribution facilities.

The development of the television system for the London Clinic provided several historic firsts. It was the first system to distribute TV signals at the broadcast frequencies over relay cable designed and intended specifically for audio distribution. In particular it was the first system to prove the feasibility of carrying TV and audio signals on the same cable pairs without mutual interference, thus making the use of existing relay cables for simultaneous transmission of both radio and television practical.

There was considerable interest and enthusiasm in Rediffusion following the successful completion of the London Clinic installation because of the possibilities it offered of providing a television relay service on a much larger scale in existing wired areas without the costs of installing coaxial cable. By this time, in 1949, Rediffusion companies throughout England and Wales had some 80,000 route miles of relay cable installed and served nearly 400,000 audio-only subscribers — a very considerable capital investment. The ability to carry TV signals on balanced-pair relay cable and combine them with audio distribution on the same pairs avoided making the existing cable obsolete, or at least duplicating it as would be the case if coaxial cable had to be used for this purpose where television services were to be added to radio services in an existing wired area. In fact it formed the basis of the system of cable television used on a large scale in Great Britain until the early 1980s when the government, as a matter of policy, decided to promote the development of multichannel cable TV systems which required the use of coaxial technology.

There was little, if any, technical advantage in using the existing relay cable for television distribution at the original broadcast frequencies. In fact the reverse was the case, since attenuation was higher than on an equivalent coaxial cable, the system was more susceptible to faults causing unbalance, and more components had to be used to maintain balanced conditions. Economy of installed cable cost was its main justification since the addition of television service at the London Clinic using the existing relay cables had cost nearly half what it would have cost had it required the installation of coaxial cable.

With the very considerable investment which Rediffusion already had in radio relay systems in many communities, the possibility of adding television distribution without making the audio distribution facilities obsolete was very attractive. There was therefore a strong incentive to investigate further these possibilities in order to take advantage of the potential economies in large-scale development.

On December 17, 1949, the BBC opened a second television transmitter to serve the Midlands, and this provided service to a number of populous areas served for radio relay by Rediffusion companies. Following this extension of the television service there was a substantial growth in the demand for rented TV sets, and with this growth came an increasing interest in the practical possibilities of distributing the signals to these sets by cable.

Although it provided a new solution to an existing problem, the London Clinic installation was basically a master antenna television (MATV) system since, like the coaxial systems in the apartment buildings, it provided an antenna service to standard receivers within a single building. The major problem in trying to extend this technology to a larger relay system was the high loss of the audio cable at TV broadcast frequencies. A potential solution lay in reducing the frequencies used for distribution on the cable. This would not only reduce the cable losses, but would have another even more important advantage when applied to an existing relay system. It could lead to technical changes in the nature of television reception which would be more in keeping with the philosophy of relay rather than that of MATV. The basic philosophy of relay was to reduce the cost of radio reception to the user by simplifying, and hence reducing the cost, of the equipment required in the home, rather than simply providing an improved antenna input to the standard off-air receiving equipment.

Since Rediffusion was now heavily into the rental of TV sets it was naturally assumed if cable distribution of television could be developed on a relay system scale, the sets, like the loudspeakers in a radio relay system, would be rented to subscribers as an integral part of the service. As such they would remain the property of the company and there was no compelling reason why they should be standard sets. It was reasoned that if the design of the receivers was changed to accept signals at a frequency equivalent to the lower intermediate frequency to which the off-air signals would be converted anyway in a standard receiver, and the signals then transmitted over the cable, this would have a two-fold advantage. First, it would eliminate the need for a tuner, radio frequency stages, and frequency changer, with their associated tubes and components, in the receiver. Second, it would achieve a substantial reduction in cable losses due to the use of lower frequencies. Furthermore, since the TV sound could be transmitted at audio frequencies and at normal relay levels on the same cable, it would be possible to eliminate all the tubes and components normally used for off-air reception of the sound, using only the loudspeaker and a volume control as in normal relay practice.

It is worth noting that the idea of a TV set specifically designed to interface with a cable system, rather than adapting standard off-air sets for this purpose, originated in England nearly fifty years ago, while a cable-compatible set only materialized on the open market in North America within the last twenty years, in spite of the extent to which cable TV was in use and the limitations the absence of such a set imposed.

These changes in receiver design resulted in substantial economies in the cost of receivers which, by comparison with the price of standard receivers, made a TV relay service potentially very attractive on the grounds of cost alone, while at the same time solving the problem of the higher losses on the relay cable. In fact it was estimated that enough could be saved in the cost of the receiver, plus a further saving in an associated antenna installation, to cover the expense of cable distribution, including rental of the receiver, at a subscriber density on the cable system of about 30 percent. This change in transmission frequency of the vision signals also had another important advantage in that it automatically overcame any possibility of direct pick-up interference, either on the cable or on the receivers, since the signals were converted out of the very high frequency (VHF) band in which they were broadcast.

This technology, based on the transmission of the visual signals at "high frequency" as distinct from "very high frequency" and the sound signals at audio, formed the original basis of the TV relay business operated by Rediffusion and similar companies in Great Britain for at least the next three decades until the mid-1980s, when the very limited channel capacity of these systems rendered them obsolescent and the principle of TV relay based on multi-pair cable was replaced by coaxial cable with its much greater capacity.

By the latter half of 1950, Rediffusion was conducting considerable development work on the practical application of the techniques pioneered in London to existing relay installations on a larger and more representative scale. This development was sufficiently advanced by the end of 1950 that it was decided to proceed with a field trial on an existing system serving the town of Margate on the coast of Kent near the Thames Estuary. Since this system was located some eighty-two air miles from the Alexandra Palace transmitter in London it was necessary to locate a suitable receiving site and also undertake a considerable series of tests on suitable antennas for long-distance reception.

By March 1951 a permanent antenna site had been established in open country on the outskirts of Margate and good reception of pictures had been demonstrated at the company's showroom in the town through nearly two miles of relay cable. This cable run included a prototype repeater amplifier designed for this application. The first subscribers were connected experimentally to the system on April 12, and the results were demonstrated to the press on the following day, creating considerable public interest. Planning then proceeded to expand the service on a commercial basis. This included the design and development of subscribers' tap-offs for connection to the cable using transformers instead of resistors to provide the required impedance match with substantially reduced losses. Simultaneously with the work in Margate a similar installation was under way on a Rediffusion system in Nottingham, receiving signals from the newly opened BBC Midlands transmitter at Sutton Coldfield, and the first paying subscribers were connected early in 1951.

Other relay companies in the U.K. were also doing development work on cable television distribution along similar lines around this time. Link Sound and Vision Ltd. was set up in 1949 as a joint venture of Pye and Murphy, two leading manufacturers of television receivers, to develop a system of television relay based on earlier experimental work. They did some further work on a relay system in Northampton, and then acquired a relay franchise in Gloucester and, using the signals received from Sutton Coldfield, commenced installation of a system there towards the end of 1950. This system was opened in March 1951 at about the same time as the commercial service in Nottingham.

EMI Ltd., another leading receiver manufacturer, also designed a television relay system, but practical trials never progressed beyond an experimental installation serving some eighteen homes in the vicinity of the EMI plant in Hayes, Middlesex, on the outskirts of London, and it was never operated on a commercial basis. Both these systems transmitted high frequency vision signals on balanced-pairs, with the sound transmitted at audio to a special wired receiver, similar in general concept to the system developed by Rediffusion; however, while the Link system used screened-pair cables, the EMI system used open wire lines.

By early 1952 the BBC had opened transmitters at Holme Moss and Pontop Pike to serve the populous areas of the north including Lancashire and Yorkshire. There was rapidly developing interest and intensive activity by the Rediffusion organization in plans for television receiver rental and wire distribution in many major centres where radio relay systems were already in operation. These included Birkenhead, Hull, Cardiff, Newcastle, Darlington, Stafford, and many other cities all served by relay systems with as many as 80,000 subscribers already connected for audio service.

By the end of 1953 conversion of portions of the audio relay service to carry television was well under way in most of these areas, and an increasing number of subscribers were receiving television service using wired receivers in conjunction with cable distribution, although many more were using rented receivers with antennas for off-air reception. At this time further work was in progress on the requirements for the transmission of two television programs on the cable system, anticipating the probable introduction of a second TV service, although this did not in fact materialize until several years later.

This work proceeded following a policy choice between three major technologies for multi-channel distribution in the context of the expected U.K. requirements: (1) VHF channels combined at their broadcast frequencies on coaxial cable — now the North American standard; (2) high-frequency channels combined on a single balanced-pair; or (3) single high-frequency channels on separate balanced-pairs. For Rediffusion, with the very heavy commitment to audio relay, the last of the three options appeared to be the most practical. The Rediffusion group did not make the final decision to complete the addition of television to the many existing radio relay systems until 1955, but thereafter conversion proceeded rapidly, using the basic technology applied on the trial system in Margate — that is, transmission of vision signals at high frequency and sound at audio over screened quad cables to wired receivers included as part of the service.

Although this TV relay technology became the standard for Rediffusion and other companies providing wired TV service in the U.K., there were parallel developments in coaxial cable technology which were in advance of any similar systems for television distribution by cable elsewhere in the world. For example, in 1951 Rediffusion designed and installed a television distribution system using coaxial cable for the Festival of Britain. This was a world fair class of exhibition built on a large site on the south bank of the River Thames in the centre of London near Waterloo Bridge. It was opened on May 4, 1951, by King George VI and ran until September 30. The television system was a complex one for the state of the art of the day as it had to provide reception of the BBC programs from Alexandra Palace and distribution, together with a number of locally originated programs, to several hundred receivers at various locations within the Festival grounds. For this purpose it used a specially designed comprehensive control centre for monitoring, program selection, and switching.

This early work on television distribution by cable also involved much experimentation with long-distance TV reception since at this time there were very few broadcasting transmitters, and the interest in TV relay was inevitably greatest where reception was difficult due to distance from the nearest transmitter. Reference has already been made to the tests conducted on suitable antennas for long-distance reception at Margate before a receiving site could be established to feed the trial system there.

About the same time, in 1950, there was considerable interest in receiving television and developing a distribution system in Jersey, one of the Channel Islands. These islands are about fifteen miles from the Normandy coast of France but are part of the British Isles, in fact the only part of the U.K. which was occupied by the Germans during World War Two.

Since Jersey is 180 air miles from the Alexandra Palace transmitter, it constituted a very real test of the techniques available for long-distance reception from well beyond the horizon and over a sea path which even today would be considered difficult. Several different types of high-gain antennas were tried in these tests, and particularly good results were obtained with broadside arrays and long tilted-wire antennas.

The tests in Jersey were not confined to reception of the signals from the Alexandra Palace transmitter in London. It was found that it was possible, with the special antennas, to receive signals from the French television transmitter which had recently been installed on the Eiffel Tower in Paris some two hundred miles away. On October 7, 1950, a successful demonstration was given to top executives of Rediffusion, the chief engineer of the BBC, and members of the Jersey States Telecommunications Committee of simultaneous reception of pictures from London and Paris. A television receiver specially imported from France was used for the latter as the French system used different transmission standards from those used by the British service. This too may have been the first time television pictures had been received off-air simultaneously from two different countries, a feature which is now commonplace across the U.S.-Canada border and many other international borders.

While these tests of long-distance TV reception were proceeding, planning was also begun to determine the method of distribution to be used. Unlike other areas on the U.K. mainland where Rediffusion was operating, there was no cable system in existence providing a radio relay service, and there was a natural reluctance to invest the considerable capital required to build a cable distribution system from scratch with the uncertainty of a technology which was still very much in the early stages of development.

The telephone system on the island was owned and operated by local government under the supervision of the Jersey States Telcommunications Committee, one of the parties sponsoring the local introduction of television. It was therefore suggested that experiments proceed with the distribution of the television signals using high-frequency carriers on the telephone lines since these already provided the required distribution facilities with access into the homes. There was some information in the technical literature on this method of distribution since there had been some experience of this technology some years before in parts of Europe.

Experimental equipment was designed and built and links were set up between the receiving site and several test sites located at hotels around the island and fed through the local telephone exchanges and the regular telephone connections. The reception of television programs from both London and Paris at these test sites was successfully demonstrated to the satisfaction of the company and the local authorities. However, at this time much was being learned in England about the susceptibility of unshielded cable to atmospheric conditions. It was realized that telephone cable, especially open-wire lines, which were prevalent on the island, would be particularly susceptible and virtually impossible to shield, and plans to use the telephone infrastructure were abandoned.

In spite of the experimental achievements in long-distance reception in Margate and Jersey, the general reception of TV programs from outside the borders of the country was not a practical feature of television reception in England for many years, indeed not until the advent of geostationary satellites effectively abolished international boundaries for this purpose many years later. In this respect the situation differed drastically from that in Canada, where, even before any Canadian TV transmitters were on the air, reception on domestic antennas was possible in a number of areas from stations across the U.S. border, and cable systems were extending the reception range considerably by using special antenna installations, thus providing or increasing program choice.

Parliament did not authorize the establishment of a second television service until 1954, and it certainly was not envisaged in 1952 that a third service, much less the proliferation of program choices made possible later by satellites, was likely to be available in the foreseeable future. Thus Rediffusion decided at that time to adapt the existing radio relay networks to carry two TV channels using balanced-pair relay cable rather than using coaxial cable which, in the thinking at that time, would have provided unnecessary excess capacity which could not justify the extra costs involved.

For more than thirty years, cable systems in the U.K. had been prevented by the terms of their licences from offering to subscribers any program material other than that available off-air from the national broadcasting system. During this time the British government pursued a regulatory policy of improving TV reception throughout the nation and increasing program choice. As a result by 1981 the broadcasting system had been expanded to cover nearly 100 percent of the country with signals of adequate strength for domestic reception, and there were three network services available providing program choice. In consequence there was little inducement to subscribe to these services through cable and the appeal of these systems fell steadily to the point where in 1981 it was estimated that within some five years most of the remaining subscribers would be lost and the cable companies would be out of business.

In February 1982 a technology advisory panel appointed by the government concluded that a sound state of the art cable distribution network would be an essential component of future communications systems, and that it was essential that such a system be in place before the advent of satellite-delivered services then being envisaged. The report made it clear that, in the opinion of the panel, unless a positive decision on the future regulatory environment for cable systems was forthcoming by the end of 1982, there was very little prospect of a modern cable industry being established in the U.K.

The government responded to this report with remarkable alacrity and in April 1982 established an "Inquiry into Cable Expansion and Broadcasting Policy" chaired by Lord Hunt of Tamworth. The inquiry, after receiving and considering 189 written submissions, produced a report with recommendations only six months later on September 28, 1982.

The Hunt Report opened with the following words: "The decisions which must be taken in relation to cable expansion are of great significance. There are no modern cable systems in this country. We have some aging narrow-band systems which do no more than relay public-service broadcasts. The new wide-band systems can provide, not only a very large number of channels, but also channels with two-way capability, allowing information to pass in both directions."

As it turned out, the future for cable TV in the U.K. was saved by this major change in the regulatory climate which favoured the installation of multi-channel coaxial systems along the lines of the systems then being built in North America. At the same time the national telephone service, which up to this time had been operated by the Post Office, was privatized and placed under a new organization, British Telecommunications. These changes, coupled with a complete reversal of the government policy on foreign ownership of cable systems, and the encouragement of cable operators to provide local telephone service on their coaxial networks, resulted in a renewed surge in cable system construction, largely managed and financed by Canadian and American interests, after some three decades of virtual stagnation.

Chapter 2

From radio relay to CATV in Canada

The development of television distribution by cable in North America was preceded by an attempt by the British company Rediffusion to repeat their successes in radio relay overseas. By 1948 Rediffusion had substantial investments in relay operations overseas, including several in the Caribbean area — Jamaica, Trinidad, and British Honduras (now Belize). With this foothold in the western hemisphere, and an interest in further overseas expansion, it was natural that the company should look to Canada as an entry point to the continental North American market, and Montreal was chosen as the location of the first Canadian radio relay system.

This choice was based on the very high housing density in Montreal, compared with any other city in Canada at that time. Montreal had a large proportion of duplexes, triplexes, and multiple dwellings, at a time when single family dwellings were the order of the day in most Canadian communities and high-rise apartment buildings almost unknown. In fact this density was closely comparable to that of most of the areas already developed for relay by Rediffusion in the U.K., where the housing generally comprised closely packed terrace homes. It could be expected therefore that the economics might be somewhat comparable. The fact that Montreal was to be the first city in Canada to receive television service, as planned by the Canadian Broadcasting Corporation, was probably not a factor in this decision, since television was a new and immature element in Rediffusion's thinking at that time, even in the U.K. where receiver rental was under way but the first TV cable distribution system capable of serving a community had not yet been developed.

During 1948 preparations were made to commence the installation of a four-program radio relay system using a double star quad network and materials and techniques similar to those in use in the U.K. A Canadian company, Rediffusion Inc., was incorporated, and construction commenced in the Lafontaine area of Montreal, an almost exclusively French-speaking area of some 88,000 households. It planned to distribute the three available French-language radio stations plus a program of closed-circuit background music as the fourth service. For this purpose an existing Muzak franchise was purchased, the existing service to commercial and industrial establishments being continued and expanded, while the programs also provided one channel of continuous music on the relay network.

At this stage Rediffusion had already sown the seeds of its ultimate failure in Canada by these early decisions. In retrospect it would appear that the company had not given sufficient consideration to the basic reason for the success of radio relay in the U.K. This was the ability to offer an inexpensive, trouble-free radio service in a climate of limited broadcast station availability and relatively high cost receivers. Circumstances in Canada, and especially in Montreal, were significantly different.

First, radio receivers were substantially less expensive and, unlike the situation in the U.K., easy credit purchase was available, making the monthly cost of owning a receiver comparable to the minimum that could be charged for a relay service by cable. Partly for this reason the marketing climate in Canada, and indeed in North America, was not conducive to rental as a cheaper alternative to purchase.

Second, on the advice of French-speaking executives of the company in Montreal, the service was concentrated in French-speaking areas of the city where it had been assumed that delivery of the three French radio stations then available, plus a channel of background music, would be an attractive service offering. It was not until after substantial capital commitments had been made, and acceptance of this service offering proved to be less than enthusiastic, that it was discovered that many of the residents in these areas listened to the English-language stations as well, both those in Montreal and those which could be received from across the border in the United States. Even without this factor, the system had been designed on U.K. lines to provide a four-program service with no provision for the later addition of further programs when available, even in French. Clearly there had been a lack of understanding of the more open broadcast climate in North America, and the competitive effects of a private enterprise element not present in the U.K., which would inevitably increase the number of stations with time.

In 1949, soon after Rediffusion began this construction program, it became known that the CBC was planning the introduction of a national television broadcasting service to commence with a transmitter in Montreal, the first in Canada, in September 1951. As it turned out the opening of the service was delayed by one year. In the light of these plans, and as a result of evidence from the United States on the public impact of television, by early 1950 Rediffusion was re-evaluating the decision to continue the development of radio relay in Montreal.

It was very evident from many sources that the impact of television on the public, wherever it had been introduced, was without precedent in any other new form of entertainment, and nowhere was this more striking than in the United States. Between 1947, when television station construction resumed after the end of the war, and the end of 1949, a total of ninety-one new transmitters had been opened for service, and in several metropolitan areas there were already four or five competing services, with six in New York and seven in Los Angeles. This growth in the number of transmitters had been matched by the increase in the number of receivers, as the following figures show: January 1946, 4,000 television sets; January 1949, 1 million; January 1950, 3.8 million; January 1951, 8.7 million.

Evidence was also available as to the effect of television on other forms of entertainment, even in those early days. A survey carried out by Warner Bros. in New York showed that over a period television set owners reduced their visits to the cinema by about 30 percent, while a similar survey in Washington showed a reduction of as much as 72 percent. Another survey showed that theatre receipts in Chicago dropped by 12 percent in 1949, by 15 percent during the six months to January 1950, and by 25 percent during January 1950. There were also clear indications in the United States, as there were in the United Kingdom, that ownership of television sets was not confined to the middle or upper income levels, even with the relatively high cost of the sets in those early days.

After studying all this evidence the management of Rediffusion took the view that there was every reason to believe that television, when introduced in Canada in September 1951, would have the same phenomenal success. They concluded that while the original plan to provide an audio relay service in Montreal was sound when it was conceived, it had been overtaken by events and an audio-only service would now be unlikely to have even a modest commercial success. Rediffusion recast the development plans for Montreal so that the prime objective would be to provide television programs when available as a form of home entertainment, at a price and of a standard of quality which would compete effectively with the off-air television receiver.

It will be noted that the idea that a cable distribution system could compete economically with off-air reception depended on the existence of one or both of two conditions: that the broadcast signals were not readily receivable in homes in the area, or that a TV relay system would reduce the overall cost to the user. The success of a TV relay system rested on the basic philosophy of relay, which included as an integral part of the service a simpler and less expensive receiver designed to be compatible with, and part of, the total distribution system. This in turn depended on the ability to market the rental of the receivers rather than their purchase. In any new venture hindsight is always so much easier than foresight, and looking back at this decision it can be seen that it served to nourish the seeds of failure that had already been sown when the original decision was made to build a four-program audio relay service in the French-speaking areas of Montreal.

The natural location for a TV transmitter was on the top of Mount Royal, a “mountain” approximately two hundred feet high in the centre of the metropolitan area, since this would permit radiation of the signal in all directions over the area to be served. However, it was believed by Rediffusion's Montreal executives that Cardinal Leger, the head of the Catholic church in Quebec and a powerful political voice in the province at that time, would not endorse the erection of a transmitting tower on Mount Royal. If this were to be the case then, whatever the alternative site might be, the mountain would be an obstruction to off-air reception in many parts of the service area, and it was probable that cable distribution might be the only way to serve these areas. Having already set up an organization in Montreal and invested heavily in an audio relay service in that city, Rediffusion's decision to extend this service to the distribution of television followed naturally. As it turned out, the CBC transmitter was installed on the top of Mount Royal removing one of the two basic conditions necessary to establish a need for wired distribution, inadequate domestic reception in part of the service area.

In hindsight, with these circumstances in mind, it is quite likely that if Rediffusion had chosen to develop a television distribution system for initial installation anywhere in Canada other than in Montreal the future history of cable TV in Canada might have been very different and in Rediffusion's favour. For example Toronto would have been a much more favourable location for such a ground-breaking venture, where there was a rapidly developing interest in television, with many TV set owners all using antennas pointing at the only transmitters then in operation across the U.S. border in Buffalo. However, Buffalo is some seventy-five miles away, and off-air reception at that distance using domestic antennas in an urban area is unreliable, so that the availability of wired distribution could have been attractive. Even so, the system Rediffusion designed for this market would not, in the long run, have been successful because it was essentially a relay system of limited channel capacity, and depended on the rental of dedicated receivers for its success.

The development of the wired television system to be installed in Montreal commenced in 1949, following the disclosure of the CBC plans for a public television service scheduled to start in 1951. In view of the results of the tests at the London Clinic in England using the existing audio relay cable for television distribution, and the fact that this type of cable was then being installed in Montreal, the possibility of using this technology was a first consideration. However, the susceptibility of an unscreened cable to the external environment, including the effects due to direct pick-up of broadcast signals in the same frequency band, made it apparent quite early that this type of cable would not be suitable in Montreal, where it could be expected that more potentially interfering signals would be present. Furthermore, the specified requirement for this system was to distribute five sound programs derived from the then available radio stations in Montreal, plus the Muzak background music service, in addition to two television programs with their accompanying sound.

With the state of the art at that time it seemed technically preferable that all programs, sound as well as vision, should be distributed as modulated carriers on coaxial cable, rather than using the U.K. technique of distributing the sound programs at audio, which would of necessity have dictated the use of balanced-pair cables providing eight separate pairs, or four cables. Since the relay principle was to be used, and this included supplying wired receivers to subscribers as an integral part of the service, there was no good reason to distribute the television signals on their original broadcast frequencies. It was decided to adopt the U.K. technique of using a lower frequency equivalent to the intermediate frequency (I.F.) receiver to take advantage of lower cable attenuation.

The design of the system was accordingly based on modulating the video components of the two television signals onto carriers at 16.0 MHz and 28.0 MHz, the former transmitting the full upper sideband 16.0 to 20.5 MHz, and the latter being inverted so that the fully-transmitted sideband of this carrier occupied the band 23.5 to 28.0 MHz. The other sideband on each channel was vestigial in accordance with the North American broadcast standards. The audio components of the two television signals plus the six sound channels were modulated onto eight carriers spaced 20 KHz apart in the band 180 to 320 KHz, using double sideband amplitude modulation.

In view of the strong probability of high power radio transmitters operating in the frequency bands selected for the sound and vision channels, it was realized that conventional copper-braided coaxial cable would be too prone to pick-up interference. This was the type of coaxial cable being used on the TV antenna installations in London, and it had been found to be subject to direct pick-up at television frequencies in the environmental exposure tests done on both quad and coaxial cables in London. It was clear that it would be necessary to use a much more efficient screen, and this virtually dictated the use of a solid metallic screen for the Montreal system. With the co-operation of a large cable manufacturer in the U.K., a coaxial cable with a solid aluminum sheath was developed specifically for this project. This was the first solid aluminum-sheathed coaxial cable ever produced and, although this is now practically the standard type of cable used on systems in North America, the cable developed by Rediffusion in the U.K. for installation in Montreal pre-dated its general use in Canada and the U.S. by more than ten years.

Two sizes of this cable were developed and manufactured for this system. One was of half-inch diameter to be used on trunks, and the second of quarter-inch diameter to be used on distribution feeders and for subscribers' drops. The same type of solid-sheathed cable was used on the drops because it was essential that the high shielding efficiency of the solid screen be carried right through to the receiver to prevent pick-up at this point in the system where cable signals are at their lowest levels. It is of interest that, after some fifty years during which a copper-braided RG59 coaxial cable has been practically the standard for subscribers' drops on North American cable TV systems, the use of a drop with a solid screen in the form of a metallic tape has now become increasingly popular for this purpose. The reason for using a solid-screened drop cable is the same as it was on that first system in Montreal — the need for greater screening efficiency against external interference when signals are present in the outside environment in the same frequency band as those used on the cable, with the additional requirement that radiation from the cable in bands used by other radio services must be minimized.

In keeping with the relay principle, special receivers were designed for use on this system, and these were known as "terminal units" to distinguish them from standard off-air receivers. Like the wired receivers later developed for the TV relay systems in the U.K., these comprised a vision I.F. amplifier switched by a program selector to accept either of the two vision channels, followed by a demodulator and video amplifier. Unlike the U.K. receivers, the sound signals were applied to a sound I.F. amplifier switched by the same program selector to accept any one of the eight sound channels, and followed by a demodulator and audio amplifier. Direct connection was made to the drop cable through a 75-ohm coaxial input, so that these were in fact the first TV receivers ever designed to be cable compatible — that is, with an input designed to match the connecting cable, and to accept the cable frequencies without any need for conversion. The first terminal units were designed and manufactured with a sixteen-inch picture tube, large for the time when the prevalent sizes available in standard receivers were nine-inch and twelve-inch, and later a 21-inch model was introduced. Fifty of these units were produced and shipped to Canada in time for the initial field trial of the system in Montreal in late 1950.

The line amplifiers originally developed for this system consisted of two types: a trunk amplifier known as a tandem repeater and a distribution amplifier known as a block repeater. Each of the amplifier stations comprised a vision repeater, consisting of a broadband amplifier covering the 14 to 30 MHz frequency band and handling the two vision channels simultaneously, and either a separate sound repeater covering the 170 to 330 KHz frequency band and handling the eight sound channels, or a by-pass filter where a sound repeater was not required. The operating output level of the vision tandem repeater was +44dbmV (44 decibels above 1 millivolt) per channel, and of the vision block repeater +48dbmV per channel, while each sound repeater operated at an output level of +46dbmV per channel. It will be noted that the vision repeaters covered a frequency band in excess of one octave, carrying two channels simultaneously, and were therefore true broadband amplifiers, and that these were developed and used in the Montreal system at a time when only amplifiers capable of handling a single channel were available to the pioneer cable systems being built at the time in North America.

Some ten miles of trunk and feeder cable with trunk and block repeaters were installed and available for initial tests by the end of 1950. These tests indicated that there was no interference with signals in the cable in the presence of strong broadcast signals in the same frequency band, thus confirming that the basic system design and the use of the solid aluminum cable were indeed satisfactory. At this time the development work on TV distribution on balanced-pair cables was under way in England. Since a considerable amount of audio relay cable had been installed in Montreal and was already in service, a major decision had to be made on the preferred television technology to be used on this system.

This uncertainty as to the practical success of either the balanced-pair or the coaxial technologies made an immediate decision on the implementation of a new plan for television in Montreal a difficult one. Consequently, it was decided to proceed in four stages: (1) to complete the field tests on the coaxial system as early as possible; (2) to continue installation of the star quad audio network in the Lafontaine area limited to about 10,000 homes; (3) in promoting the radio relay service to promise to include television when it became available; and (4) if the field trial proved this technology to be superior to balanced-pair distribution in the Montreal environment, then installation of a coaxial network on a commercial basis should proceed not later than March 1951, so that this promise could be fulfilled by the time the broadcast television service started as then expected that September.

Rediffusion had serious doubts by now whether a service consisting of sound programs only, that is a radio relay service, could be successfully marketed in Montreal in view of the anticipated public demand for television when it became available. The company felt that the earlier introduction of a radio relay service to a limited area could be justified if it was linked with a definite undertaking to include television when available, and this would at the same time provide very useful operational experience. Its main value in the company's view was to promote among the public some knowledge and appreciation of the advantages of wired distribution of programs and, more important, it would provide the opportunity to promote the idea of receiving television by wire.

By the spring of 1951 the coaxial field trial had been successfully completed and had proved to be superior to the balanced-pair technology, at least in the North American environment. The decision was made to proceed with construction of a coaxial network on as large a scale as possible to be ready for the expected opening of the CBC television service later that year. In fact, the start of this service was delayed for a year. By the time the first station, CBFT, went on the air on September 6, 1952, more than two hundred miles of coaxial cable had been installed, passing 58,000 homes. In addition, by this time there were 275 miles of star quad network passing 75,000 homes with four-program audio service. Approximately half of this network was duplicated — that is, coaxial cable and star quad cable running side by side — so that there were in fact about 94,000 homes to which a wired service of either television or audio or both was available.

In 1951, when the first subscribers were connected to the coaxial system as part of the field trial, there were no broadcast TV programs available due to the delay in the CBC plans. A studio was installed, equipped with a film chain and slide projector, together with two cameras and equipment suitable for originating live material such as news and commentaries, and these provided a daily television program to Rediffusion subscribers for at least the next year. A considerable subscriber total was built up using mainly French-language films and "talking head" type programs. This was probably the first operating TV-origination studio in Canada, and certainly the first cable TV system in North America to originate its own programming on a regular basis, nearly eighteen years before the Canadian Radio-Television Commission (CRTC) made local programming a mandatory requirement on most cable systems in Canada.

When CBFT began broadcasting on channel 2 on September 6, 1952, it was carried on one channel of the two-channel system and the local originations were continued on the second channel. This offering, with the program choice not available to owners of standard receivers using off-air reception of the CBC station, proved to be very popular, particularly as television entertainment of any sort was a novelty. By the end of 1953 the number of subscribers to the TV relay system had increased to nearly 6,000. Then on January 10, 1954, the CBC opened a second station in Montreal, CBMT, broadcasting also from Mount Royal on channel 6. Since this station was readily receivable by every domestic antenna the program choice advantage of the cable system was immediately lost, and this was aggravated by the subsequent actions of the CBC.

At this time the CBC Board of Governors was the licencing authority for all broadcasting in Canada. Although the newly developed system of wire distribution did not fall under its mandate, the board was strongly opposed to what it viewed as competition in broadcasting anywhere in Canada, and particularly in Montreal. When CBMT went into service the board lost no time in bringing political pressure to bear on Rediffusion to add it to the cable service, so that both CBC stations were carried, even though CBMT was an English-language station and the majority of the cable subscribers were French-speaking. Submitting to the pressure, which included clear threats that if Rediffusion did not comply the system would be closed down, CBMT was added on the second channel, while a limited amount of local origination, mainly in French, was continued outside the then limited CBC broadcasting hours. However, this was not sufficient to counter the availability of both CBC stations off-air using standard receivers and "rabbit-ear" antennas, or the tendency of the public to want to purchase their own receivers rather than rent, and the number of subscribers started to drop off.

Within a year or so after the second CBC television station in Montreal went into operation, three TV transmitters started service in the United States across the nearby border and within reception distance of Montreal. These were WCAX-TV in Burlington, Vermont, broadcasting on channel 3; WPTZ in Plattsburg, New York, on channel 5, and WMTW-TV located on the top of Mount Washington in New Hampshire and broadcasting on channel 8. Although within nominal reception distance of Montreal, these were all distant stations and could not be received on simple rabbit-ear antennas as could the local stations. In addition to the effects of distance, the first two stations were even more difficult to receive because of interference from the two CBC transmitters on the adjacent channels 2 and 6, and the blocking effect of Mount Royal for many areas of the city. There was clearly a public interest in receiving these U.S. stations, not only for the increase in program choice, but also for the already evident curiosity in TV programs being produced in the United States as an alternative, or at least an adjunct, to the CBC productions.

It was this development, perhaps more than anything else, which finally brought Rediffusion to the realization that the broadcasting climate in North America differed in several fundamental respects from that in the U.K., which had been the basis of their previous experience and their success. First was the basic antipathy to rental in a climate in which, unlike the U.K., it was common for appliances such as television sets to be purchased with instalment payments, and these payments often compared favourably with the cost to lease or rent a relay receiver. Second, in Canada family mobility was much more a fact of life than it was in those areas in the U.K. in which the company had been most successful, so a television set that could only be used in conjunction with a compatible cable system was a disadvantage whether owned or rented. Third, the early potential for access to a multiplicity of broadcast transmitters, both radio and television, had not been considered in the design of either the radio or the coaxial TV relay systems, and this inevitably left the systems with inadequate channel capacity, and very little technical flexibility for economically enlarging this capacity, to meet increasing market demands.

By 1955 these problems became increasingly evident from the number of television sets of standard design which were operating in homes served by the Rediffusion network, while the number of subscribers to the relay service steadily decreased. It was clear that the market for rental of the terminal units was becoming severely restricted, while at the same time there was a growing need for cable distribution to standard receivers as an alternative to off-air reception, particularly of the newly available distant U.S. stations. Accordingly, it was decided to add a third TV channel to the system, and at the same time to add FM sound carriers to the three vision channels. A frequency converter was then designed capable of converting these vision and sound carriers to a channel in the VHF television band for input into a standard receiver.

The availability of subscribers' converters potentially widened the market for the cable service by removing the need to rent a special receiver, making this service available to the owners of standard sets; however, the converter impact on the business was limited, so long as the cable system itself was limited to three channels and could offer only the two local stations and one of the three U.S. stations then available. Rediffusion then commenced to modify all the terminal units by the addition of a tuner and FM sound section to convert them, in effect, into standard sets capable of receiving the local stations off-air. By the time this program had been completed in 1958 all the installed terminal units had been supplied with rabbit-ear antennas for local reception, the local stations had been removed from the cable, and the distant stations substituted. At the same time the distribution network was expanded to serve additional areas located behind Mount Royal, where reception of the distant stations was impractical and which, being predominantly English-speaking, offered a better market for U.S. programs. These programs were then available to subscribers who already owned their own sets by adding the converters at the set inputs to access the cable channels.

Thus were created two more firsts in cable TV technology — the introduction of a set-top converter (in this case to overcome the limited channel capacity of the cable system) nearly twenty years before a converter similar in principle was introduced to overcome the limitations of the standard twelve-channel receiver; and a modified terminal unit to make it off-air compatible, while standard receivers modified to be cable compatible were still in the minority.

These changes in effect converted the television relay system in Montreal into the equivalent of a Community Antenna Television (CATV) system and sounded the death knell of the relay principle, at least in North America. The progressive development of this principle, as applied to television reception first in England, had produced a number of technical innovations which were used, sometimes in different form, in later cable TV systems. But TV relay, as distinct from CATV or cable TV as it later became known, succumbed to the prevalent desire of the viewing public in North America to own their own receivers rather than rent, a desire made much easier than in the U.K. by the climate of easy credit.

By 1959 the coaxial cable network in Montreal had been extended to cover 160,000 homes with some six hundred route miles of cable, making this the largest CATV system at that time in North America, and indeed in the world, and the first to serve a large metropolitan market. There were some 14,000 subscribers connected to the system using their own receivers with converters, and some 2,000 still renting the Rediffusion terminal units suitable for use only on this system.

Chapter 3

The advent of CATV in the United States

Television broadcasting did not start in the United States until 1940, nearly four years after the BBC Alexandra Palace transmitter opened for public service in England. By that time the BBC service had closed down temporarily with the outbreak of war in Europe.

The first transmitter to go on the air in the U.S. was WBBM-TV in Chicago, which opened in August 1940 with the call sign W9XBK. It was almost another year before two more stations opened in New York on July 1, 1941, WNBC-TV and WCBS-TV, as flagship stations of the networks whose initials were incorporated in their call signs. After that only three more stations opened during the next five to six years: KYW-TV Philadelphia in September 1941, WLS-TV Chicago in October 1943, and WNEW-TV New York in May 1944. Due to the American war effort there were no new television stations licenced between May 1944 and January 1947.

With the end of World War Two in 1945, the American economy gradually returned to a more normal civilian footing, and plans were soon under way to resume the interrupted development of television, the newest entertainment medium. In 1947 twelve new stations were licenced, and in 1948 a further thirty started service so that by the end of that year there were forty-eight stations in operation across the U.S.

The receiver manufacturers were also doing their best to keep up with the public demand generated by the rapidly increasing television broadcasting coverage. In 1946, with six stations broadcasting, there were only some 5,000 sets in use. By the end of 1948 this figure had increased to nearly one million. All the transmitters were located in major cities and metropolitan areas having very large potential markets within easy reach of their broadcast signals, but with the accelerating growth in stations it would not have been too long before the many smaller cities and towns would also have been served by further broadcast transmitters. However, this growth was suddenly brought to a halt in 1948 when the broadcast licencing authority, the Federal Communications Commission (FCC), imposed a freeze on the issuing of new TV station licences.

When the FCC adopted the first rules for the licencing of television broadcasting in the 1930s practical experience with radio transmission in the VHF band, which was to be used for television, was very limited, and it was not expected that these broadcast signals would be propagated beyond the horizon. This implied that reception would be limited to locations within line of sight of the transmitter. After public television broadcasting started, and practical experience of reception under various conditions and at various distances from the transmitters accumulated, it became apparent that this assumption was not in fact correct. Frequent reports were received of TV reception from transmitters well beyond the horizon, although the quality of the reception was not reliable since it was apparent that it depended on the nature of the intervening terrain, meteorological conditions, and other more or less complex factors.

On the face of it, this long distance reception might have been an advantage, extending the potential coverage area of a transmitter quite considerably, if it had not been for another unforeseen circumstance. The allowable distance between transmitters on the same channel or on adjacent channels had been formulated on the assumption that the broadcast signals stopped at the horizon. When it was found that they did not it became clear that some of the spacings might be too close, giving rise to substantial interference between stations when receiving in the fringe areas between them. This situation called for a major study and rethinking of the rules, and the FCC imposed the freeze to provide the necessary time and opportunity to make any resulting changes before too many commitments had been made with new stations.

A further forty-nine transmitters started operation in 1949, having received their construction permits before the freeze, but it had its effect because only nine stations started in 1950, one in 1951, and nine in 1952. After new licencing rules were adopted the freeze was lifted in 1952, and 156 new stations went into service in 1953. However the freeze had a serious, if temporary, effect on the growth of television service.

The major effect of this pause in growth was a delay in the extension of television service to many smaller cities and towns where considerable interest had been generated by media reports from the areas where it was already available. A by-product of this delay was that many appliance retailers in these unserved areas in anticipation of television coming to their markets had laid in substantial stocks of TV sets to meet the demand, only to see the market dry up when the licencing freeze shelved plans for an indefinite period. Even in some communities where it had been expected that television might be receivable from the new transmitters, it was found from experience that many homes were unable to receive it because mountains or other obstructions blocked the signal. However, as the saying goes, there is more than one way to skin a cat, and this situation was a challenge to several entrepreneurial spirits who were determined to do something about it.

Probably the first man in the United States to develop what was eventually to be a Community Antenna Television (CATV) system was John Walson in Mahanoy City, Pennsylvania, who had an interest in an appliance store. By 1948 there were three telelevision transmitters operating in Philadelphia, the major metropolitan area some eighty miles from Mahanoy City, and there were substantial expectations that signals from these transmitters could be received at this distance. To cater to this anticipated demand John Walson began selling TV sets in his store. But in spite of the public interest in this new entertainment medium the demand for these sets was severely restricted because of difficulty in receiving the signals, as the community is in a valley between mountain ranges.

The location of Mahanoy City is typical of many communities in the state of Pennsylvania, which is characterized by numerous mountain ridges running roughly from northeast to southwest. Naturally, the roads tend to run along the valleys between the ridges, and the towns are located in these valleys. Consequently, in this typical community the majority of the homes were shadowed from Philadelphia by the mountains, and off-air reception on domestic antennas was impractical. This same problem was encountered in the appliance store where Walson wanted to demonstrate his TV sets so he could sell them. He overcame the problem in the store by building an antenna on one of the nearby mountains and running army surplus twin cable down the mountain from tree to tree to the store, where he was able to produce reasonable pictures from each of the three Philadelphia stations.

This of course generated considerable interest in television in the town, and produced a potential market for the sets he had in stock. However, pictures in the store would only sell sets if those same pictures could be produced in the homes of his customers, and Walson then set about doing just that. He obtained permission from the Pennsylvania Power and Light Company to attach his wire to some of their poles and then proceeded to extend his twin cable to anyone he could reach from the store who would buy a television set from him. During this period in 1948 the installations were on an experimental basis and he did not charge for the service.

Right from the start he had some problems with varying signal levels which were being affected by weather conditions, and it became clear that twin cable was too vulnerable to environmental influences. In fact Walson was having exactly the same problems as those being experienced at the same time by Rediffusion in England using unscreened audio relay cable. So later in 1948, he replaced all the existing twin cable with army surplus coaxial cable and by 1949 started to operate a commercial service, charging subscribers $100 for installation and $2 per month.

In the following year, 1950, Walson's success attracted the attention of another man in Mahanoy City. William McLaughlin decided to build a system starting from the other end of the town, so as not to duplicate what Walson was already doing. McLaughlin was assisted by a small manufacturer of electronic equipment based in Philadelphia, Jerrold Electronics Corp., owned by Milton Jerrold Shapp. He had developed, and was manufacturing, equipment for use by dealers to demonstrate and sell their sets, and earlier that year had successfully completed a TV antenna distribution system in an apartment building in New Jersey using this equipment.

The amplifiers supplied by Jerrold Electronics and used in these installations were sold under the trade name Mul-TV. These were single channel strips linked together to handle a total of three non-adjacent channels in the low band, for example 2, 4, and 6. They were not suitable for cascading — using in series — and were essentially what would be known today as MATV amplifiers. Nevertheless, Shapp could see the possibilities in the potentially wider application to community distribution systems and gladly cooperated with McLaughlin in his venture. Twenty years later, in 1970, this system was purchased by Walson and the two were merged into one.

At about the same time as these systems were being constructed in Mahanoy City, Bob Tarlton put together a group to build a system in Lansford, Pennsylvania. Tarlton started his system in a somewhat more sophisticated way than Walson because he was able to negotiate formal pole attachment rights with the Pennsylvania Power and Light Company and with the local telephone company at a charge of $1.50 per pole per year, probably the first such pole attachment contract for cable TV. He also used coaxial cable from the start rather than twin. The antennas for this system were mounted on an eighty-five foot tower on top of a hill three miles out of Lansford, and approximately seventy miles from the Philadelphia stations.

As a result of Shapp's forward looking policy, Jerrold Electronics also cooperated in the Lansford project. Although Mul-TV amplifiers had been supplied to McLaughlin to build his system, Shapp was not satisfied that the amplifiers were suitable for this purpose and, as an engineer, he recognized only too well their inadequacies when cascaded in a longer cable run. These amplifiers had originally been designed as boosters, or antenna amplifiers, intended to improve the reception on domestic insallations where signals could be received from a transmitter but the distance was too great for adequate or reliable reception. As such they were adaptable to dealer installations for demonstration purposes, or to apartment installations, as Jerrold had proved in the New Jersey project, but were not suited to the requirements of a more extensive community distribution system where the greater lengths of cable necessitated operating a number of amplifiers in series.

When Bob Tarlton started to build his system in Lansford and asked Jerrold to supply the equipment, Shapp saw the possibilities and agreed to provide the Mul-TV amplifiers if his engineers could experiment on the system to determine what design changes would be required to make them more suitable for this type of application. By the middle of 1951 Jerrold had incorporated design changes and brought out a new line of amplifiers more suitable for use in CATV systems.

At this point Jerrold decided to set a policy guiding the sale of this equipment since it had had some awkward experiences. Jerrold had been selling the apartment house equipment through distributors on the understanding that it was not to be used for CATV systems, in which there was a growing interest. However, some distributors were selling the amplifiers for this purpose without any mention of their technical limitations and without any engineering guidance or know-how. Other purchasers were buying the amplifiers from distributors as apartment house equipment and then attempting to use them, with limited success, for community distribution applications. In consequence Jerrold was receiving an increasing number of calls to service equipment which was not functioning as expected by the purchasers.

With the introduction of the new line of amplifiers designed to be more suitable for CATV applications, Jerrold decided to sell the equipment only in conjunction with a field engineering contract. The company would be responsible for the design and installation of the system and could therefore ensure that the equipment would perform as intended by the designers and as expected by the puchaser. Thus was started the so-called "turn-key contract" type of construction which has been applied to many cable TV systems in the United States right up to the present day.

The system in Lansford was in service by the end of 1950, charging $100 for installation and $3 per month, for three channels. It attracted considerable publicity, not only for Tarlton and his company, Panther Valley Television, but also for Jerrold Electronics which by this time was hard pressed to keep up with the growing demand for its equipment, partly because it was still a small company with limited resources and partly because of continuing shortages of material in the post-war period.

This publicity attracted the attention of some of the larger electronics companies. Philco Corp. had already realized the potential of CATV as an aid to selling television sets and arranged to sell Jerrold's equipment through their own distributors and, following the early success of the Lansford system, RCA began to take an interest. One of RCA's dealers was Martin Malarkey who owned several music stores in the area of Pottsville, Pennsylvania, about the same distance from the transmitters in Philadelphia as Mahanoy City. Like John Walson, he had a considerable number of television sets which he was having difficulty selling because of poor reception in and around Pottsville.

Malarkey could see the need for, and the opportunities in, television distribution by cable from a well-placed antenna, and he arranged with RCA to use their equipment to experiment with the idea. He commenced erection of coaxial cable in Pottsville about the time Tarlton was doing his construction in Lansford, and connected his first subscriber soon after Tarlton did in early 1951, charging $135 for installation and $3.75 per month. Later that year he acquired a small Dage camera and occasionally produced some local programs. This was probably the first cable system in the United States to originate closed-circuit programs into private homes, although not the first in North America, since similar equipment had been used on the system in Montreal a year earlier.

The cable system which was officially credited with being the first in the United States is one in Astoria, Oregon. "Officially" because there is a granite monument there, erected in 1968 by the owners at that time, Cox Cablevision Corp., on which is inscribed: "Site of the first community antenna television installation in the United States completed February 1949, Astoria, Oregon. Cable television was invented and developed by L.E. (Ed) Parsons on Thanksgiving Day, 1948. The system carried the first TV transmission by KRSC-TV, Channel 5, Seattle. This marked the beginning of cable TV."

Without in any way belittling what Parsons did as one of the early pioneers of the industry, the wording of the inscription would seem to make somewhat exaggerated claims in the light of the history that has already been recorded. Clearly he did not "invent" cable television, because in fact the distribution of a single channel over coaxial cable had been achieved as a commercial service in England twelve years earlier in 1936, and by 1949 when the system in Astoria was built there were more than one hundred small coaxial systems in service in London, England. It is true that each of these systems did not extend beyond a single amplifier and were not cable systems in the sense, or to the extent, that we know them today, but they did incorporate all the elements of such a system — wideband amplifier, coaxial cable, tree structure, impedance-matching tap-off units, terminated line, etc., and as such were basic cable TV systems.

There is also some uncertainty as to the claim that the Astoria system was the first in the United States, particularly in the light of the history of John Walson's development in Mahanoy City, Pennsylvania. Walson was experimenting and building his system during 1948, and although he was not operating a commercial service making charges to his subscribers until some time in 1949, it is likely he had homes connected to the system prior to November, 1948. Furthermore, Parsons could not have been doing much actual experimenting before Thanksgiving Day, November 25, 1948, because that was the date on which KRSC-TV Seattle (now broadcasting with the call sign KING-TV) first went on the air, and there was no other station at that time that could have been received in Astoria which Parsons could have used for his experimental testing.

In spite of these comments Ed Parsons must be credited with being one of the earliest pioneers of cable television, particularly as he went one important step further than John Walson, Bob Tarlton, or Marty Malarkey. Each of these pioneers had to overcome the shadow effect of mountains to receive television from transmitters some seventy to eighty miles away. And each built an antenna system on a nearby mountain top where good reception was available and then extended this down into the adjacent community by cable. Parsons could not do this since any nearby mountain top was still 125 miles from the transmitter in Seattle and therefore still beyond the horizon across intervening mountains.

What Parsons did do was to discover, as others in similar locations later discovered, that in mountainous country one could take advantage of the very mountains which caused TV shadows by finding isolated spots where reception occurred due to the bending of the signals by those same mountains. This phenomenon is now well known as "knife-edge diffraction" but in those days it was something new, since even the horizon had been considered an effective barrier to TV signals. If a mountain peak or a sharp ridge resembling a knife edge can be seen from both the transmitter and the receiving antenna then the signal will travel straight to that peak, be diffracted, or bent, over it, and travel on to the receiver at the other side. If the peak is high enough the distance between transmitter and receiver could be well in excess of one hundred miles. Furthermore, if the peak is much higher than the intervening terrain then the signal will travel through the atmosphere well clear of the ground and is not subject to the loss normally incurred by a signal travelling close to and parallel with the ground. Under these conditions a much stronger signal is received than would be possible by direct propagation, and the signal is much less subject to fading than most distant signals would be.

The key to this anomalous reception was to be able to locate an antenna at one of the very few spots at which this occurred. Ed Parsons did not have to search for such a location, but discovered quite by accident that he could receive the signals from the Seattle transmitter, 130 miles away, almost literally in his own back yard in Astoria where he did his initial experimenting. Later he built an antenna on the roof of the John Jacob Astor Hotel, a short distance away in the centre of Astoria where reception was also possible, and from this location he ran a cable, first to his own home, then to a receiver in the hotel lobby, and to a music store across the street, and later to other residents in the town.

At the time he started to take an interest in long distance reception of television, Parsons was operating a radio station, KAST in Astoria. Even as he was building the cable system in 1949 he did not see himself primarily as a cable operator. In fact he had experimented with an ultra high frequency (UHF) broadcast transmitter and had prepared an application to the Federal Communications Commission for a licence to rebroadcast the TV signals from Seattle if the results of his experimental reception proved adequate. His early success with distribution by cable prompted him to continue with this medium and drop the rebroadcast idea which involved the procedures and inevitable delays awaiting FCC approval. Later, after he had developed the amplifiers and other equipment he used on the cable system, he started selling them to others for installation on similar systems, and also offered a consulting and sales service, advising on the techniques he had used for cable distribution and on problems of reception. In fact there is little doubt that he was the first cable television consultant.

The pioneers of cable television in the early 1950s were not large, well-financed corporations. They were small-town entrepreneurs who could see a need and a public demand growing out of a new entertainment medium which was attracting considerable publicity but was being denied to the many living beyond the reach of the few metropolitan area transmitters which had been built before the FCC-imposed freeze on transmitter construction.

It is well known today that cable TV is a capital-intensive business requiring considerable investment up front to build the antenna system and the distribution network before any subscribers can be connected and revenue starts to flow. But in those early days it was unknown and unproven as a business, so it was virtually impossible to interest any bank or lending institution in providing the considerable financing required. It was for this reason that the early systems charged substantial fees, of the order of $100 to $150, for connection. It was essential to get their money up front because this was the only way they could finance continuing construction. These high installation charges were common for at least the next ten or fifteen years, until the industry had become sufficiently established and soundly based that it was able to attract institutional financing.

A major concern too, which made external financing even more difficult, was the possible effect that the eventual lifting of the FCC freeze on transmitter licences might have on existing or future cable systems. It was generally supposed that the whole country would then be progressively covered by television transmitters, and that unserved areas or reception from distant transmitters would become a thing of the past, making cable distribution unnecessary.

It was not until well after the freeze was lifted in 1952 that it became obvious that the cost of building and operating a television station was such that every small or medium sized community would not have its own station, at least for many years to come. Like the programs they carried, which were designed for mass appeal and therefore mass audiences, the transmitters would inevitably be built first in the larger centres where maximum market coverage existed. This is why cable television developed in the United States primarily as a small town service. Only in the late 1970s and early '80s, with the development of much greater sophistication and the addition of other services and program sources, did it penetrate into the metropolitan areas where broadcast television was first available.

In fact it was the availability of satellites for national distribution of television services in the late 1970s which changed the whole picture. Until that time the FCC, which is the U.S. regulatory agency (equivalent to the CRTC in Canada), had refused to approve the carriage on any cable system of "distant stations," defined as stations not licenced to provide broadcast coverage of the area served by the cable system. Since all metropolitan areas and most cities of any size were served by each of the three TV broadcast networks, this did not in most cases leave much scope for adding to the program choices available off-air, and was not much of an inducement to the construction of expensive cable systems.

The advent of television distribution by satellites, commencing with Home Box Office (HBO) in 1975, completely altered this situation. Now these services could be received literally anywhere the satellites could be seen, and they were no longer "distant stations," so the FCC rule became meaningless. Freed from this regulatory restriction, not only were cable systems able to receive and distribute these services, but the choice of programming offered, coupled with the cost and complexity of domestic satellite reception, substantially increased the attractiveness of cable service, and gave the industry the momentum it needed to expand into hitherto unserved metropolitan areas.

CHAPTER 4

Early CATV developments in Canada

Prior to 1952, when the first television transmitter commenced operation in Montreal, there was no Canadian public television service. But this did not mean that television was not available in Canada. Although the FCC did not lift the freeze on new TV broadcast licences in the United States until 1952, several transmitters had been built in U.S. cities close to the Canadian border before the freeze halted construction in 1949. These were located in Detroit, Cleveland, and Buffalo, each close enough to one or more major Canadian cities and other communities along the border that some reception was possible with a good antenna in a favourable location. Thus, reception was easily available in Windsor from Detroit, possible in the Toronto area from Buffalo, and also possible in London from Cleveland, although more distant and therefore more unreliable.

As a result there was a trickle of purchases of television sets in these areas. Only a trickle because few Canadians were close enough to these stations to obtain reliable reception using the relatively simple antennas then available for home installation, or were prepared to invest in more elaborate installations with little assurance of satisfactory results. Nevertheless, the potential availability of these transmissions was an inducement to several entrepreneurs to devise the means of improving their reception and so bringing television entertainment, first to their families, then to their neighbours, and later to their communities. These first efforts in Canada were just as entrepreneurial as those taking place in the United States at about the same time, but for quite different reasons. Whereas the latter were motivated by dealers anxious to promote a market for their unsold TV sets caught by the FCC freeze on new transmitters, Canadians near the U.S. border were literally reaching out for this new medium simply because it was there.

One of the first of these was Ed Jarmain, living in London, Ontario, and his early efforts illustrate this comparison between the motivations in Canada and those in the United States. Jarmain was involved in a dry cleaning business in London, but as a hobby he had always been interested in amateur radio. Around 1950, he was experimenting with antennas for the reception of FM radio which was just starting. While experimenting in the VHF band, which includes FM radio, he discovered the presence of television signals and turned his attention to the problems involved in receiving them.

The TV signals were from Cleveland and Detroit, each about 110 miles from London and well beyond the horizon, which was still thought to be the practical limit for television reception. The only antennas then available for home TV reception were simple types known as yagis with three, or at the most four, elements, and at that distance reception with these was very chancy. Sometimes a recognizable picture would be received, but more often than not there was lots of noise and static, and if motion could be detected on the screen at least you knew you were tuned to a station.

After reading up on antennas, Ed Jarmain's attention turned to the rhombic, a long-wire type of antenna, normally strung from four masts spaced in the form of a diamond, with the long axis of the diamond pointing towards the station to be received. His backyard was not large enough to accommodate such an antenna with the supporting guys required on each of the masts at the corners, but he persuaded a neighbour to allow several of the guys to be anchored on his property. The neighbour cooperated because he happened to be a Canadian distributor for Philco and had a TV set in his home on which he had been endeavouring to obtain reception from the distant stations. (Philco was one of the American receiver manufacturers who had developed an interest in CATV from the success in Lansford, Pennsylvania.) The long dimension of Jarmain's property determined the orientation of the rhombic antenna, and since this was in the direction of Cleveland its stations were the ones he received rather than those in Detroit which was off to the side of the antenna. This antenna gave substantially improved reception and generated considerable interest among friends and neighbours.

In 1951 Harry Anderson, who was in the car radio sales and repair business in London, heard about the antenna and started to pay frequent visits. About this time an article appeared in one of the trade magazines describing the cable system that Marty Malarkey was building in Pottsville. It so happened that an RCA sales representative who had supplied the equipment for Pottsville heard of Anderson's interest in long-distance TV reception and got talking to him, presumably with a view to finding a new market for the equipment which

RCA was producing for the early systems in Pennsylvania. In turn Anderson talked to Jarmain, and they decided to visit Pottsville with the RCA representative and see the system for themselves.

By this time, Malarkey had some 1,800 subscribers and the system was running on a commercial basis. He had acquired sufficient experience to be able to assist Jarmain and Anderson with advice as to how to start a system and some practical information on the costs involved and the financing required. They were very impressed with what they had seen in Pottsville and decided to proceed with a similar system in London; however, problems arose in arranging a business partnership between them and eventually they each proceeded independently.

Ed Jarmain arranged a partnership with his neighbour and they built a larger and more efficient rhombic antenna on their two properties, with a view to using it as the nucleus of a cable system if the results proved to be sufficiently good and consistent. Construction of this antenna was completed in August 1952. Although they were satisfied with the results, they were playing it cautiously and decided they needed a market trial to test public acceptance before making the much larger investment necessary for a commercial cable system.

Jarmain had already approached Ontario Hydro and Bell Canada for permission to use their poles and been turned down by both utilities. So he approached his immediate neighbours with the proposition that he would extend the television programs received on the antenna to them on an experimental and non-paying basis to the end of the year if they would report the results regularly. Of fifteen neighbours approached only two had television sets, so Jarmain arranged with Philco to lend sets to the other thirteen for the duration of the test.

The understanding with the neighbours was that if they found the service still acceptable by the end of December and agreed to continue as subscribers then they would pay an installation charge of $150 and a service charge of $4 per month. By that time, out of the fifteen participants, one had moved away, one declined the service, and the remaining thirteen retained it and paid as agreed — a very encouraging result, and probably the first commercial CATV system in Canada.

In spite of these early results, the system was not expanded at this time because of three coincident events. First, at the end of October, right in the middle of the field trial, Jarmain's neighbour died of a heart attack, and this left him without a partner and with insufficient financial support. Then, towards the end of December, news was released that a local television station was to be built in London and was expected to be on the air within the next year. Finally, before the cable could be extended it was necessary to obtain permission for pole attachments, and Bell offered to erect the cable and lease it at $1.25 per hundred feet. In Jarmain's opinion these charges were exhorbitant, as indeed they later proved to be when they were reduced to twenty cents per one hundred feet. However, Harry Anderson, who was proceeding with his own system, disposed of any chance of negotiating these rates by accepting them as he was impatient with the delays and anxious to start construction.

This was the start of Bell's policy towards CATV which was one of reluctant tolerance, based on their refusal to allow any "foreign" wires or equipment on their poles. Bell justified this as a means of protecting its own lines and equipment from possible damage if "foreign" workmen were allowed access to the poles. Bell rigidly applied this policy for at least the next thirty years, until the early 1980s, when it was relaxed to allow cable systems attachment rights under controlled conditions.

When Jarmain considered these three factors — loss of a partner with strong financial support, the high cost of installing cable if Bell poles had to be used, and the advent shortly of a local TV station receivable off-air — he decided to delay extending the system. In the meantime, he continued the service to his thirteen subscribing neighbours, and from 1952 to 1958 this number gradually increased to about thirty-two by extending the cable to others in the vicinity not needing poles to reach them.

At this time the system was carrying three channels (2, 4, and 8 from Cleveland) receivable on the rhombic antenna, all distributed on the channels on which they were received using single channel strip amplifiers at the headend, and SKL212 broadband distributed amplifiers where needed on the line. This amplifier was probably one of the first true broadband amplifiers to be introduced for use on the early CATV systems. Designed by Spencer-Kennedy Laboratories in the United States, it used a distributed line principle in which several vacuum tubes operated in parallel rather than in series, providing a fairly low but acceptable gain over a relatively broad band at the television frequencies, and was thus able to handle several channels simultaneously. Although the distributed line principle was not used in later amplifiers, the SKL212 was undoubtedly the predecessor of the type of amplifier now generally used in all cable TV systems and which has now been extended in bandwidth far beyond the capability of those early amplifiers.

While Ed Jarmain was taking a cautious approach and delaying further expansion of his service, Harry Anderson had gone ahead, building a system in another part of London and by 1958 he had between 500 and 1,000 subscribers connected. Early in 1958 Jarmain was receiving a number of calls from people, not in his immediate neighbourhood, who had heard about his antenna and were interested in having the service he was providing.

The local station, CFPL-TV, had started service in November 1953, but was not carried on the cable system because the transmitter was only a mile down the road and, far from arranging for subscribers to receive it through the cable, it was impossible to keep it out! The signal in the area was so strong that no antenna was required; it was picked up on any receiver without the need for either cable or an antenna. The calls for cable service were presumably from people for whom the novelty of the one local station had worn off, and they were interested in the wider choice of programs the cable system could offer. As a result, Jarmain decided that the time had come to proceed with his original plans for a more extensive system. By 1958, negotiations between Bell and the newly formed National Community Antenna Television Association of Canada had resulted in a cable leasing agreement at much more reasonable rates than had been previously offered. Jarmain then proceeded with plans to wire a neighbourhood of five hundred homes adjacent to the small area near his home he had been serving for the past six years.

Around this time Famous Players Canadian Corp., based in Toronto, was planning to wire an area of some 5,000 homes in North London for an experimental pay-TV system by Telemeter (as described in Chapter 5). Famous Players had made substantial commitments, particularly to Bell for cabling, when it decided to move the pay-TV field trial from London to Etobicoke, a suburb of Toronto. Rather than back off on these commitments, it decided to proceed in London with a CATV system.

At this time, while his extension was still in the planning stage, Ed Jarmain was asked to remove the rhombic antenna from his neighbour's property. This necessitated a search for an alternative antenna site, both to continue the service to his original subscribers, and to serve the new area. Famous Players, which was also looking for a suitable antenna site for its proposed system, heard of Jarmain's problem and suggested that they cooperate in finding a location for a headend which both could use. Having found a suitable site, negotiations for a shared headend progressed very quickly to a full partnership between them in the cable system, and London TV Cable Ltd. was incorporated on September 1, 1959, with Jarmain and Famous Players as equal partners. Construction of an expanded system by the new company started early in 1960. In 1970 this partnership was combined with other companies in which Famous Players had an equity interest to become Canadian Cable Systems Ltd., in order to meet government mandated Canadian ownership requirements. It in turn later became part of the Rogers Cable TV conglomerate.

While Ed Jarmain was building the big rhombic antenna in his backyard and preparing to wire some of his neighbours for a field trial in the fall of 1952, another entrepreneur in Guelph, Ontario, sixty miles from London, was developing an interest in the possibilities of CATV. Fred Metcalf was co-owner of radio station CJOY in Guelph, and Jake Milligan was his station engineer. Fred handled the sales and business side of the radio station, and in 1952 he was looking for ways to expand the business. Curiously, the thing that drew his attention to CATV was a memo from the Canadian Association of Broadcasters saying that, since licences for private television stations were not yet being considered in Canada, members might be interested in this new medium. This sounded interesting to him, particularly when he learned that a system was being built in Burlington, Vermont, not far from Montreal, where he had to visit radio sales agencies. Following this visit he went on to Burlington and talked to the owner of the system, Dr. Abajian, who, as a radiologist at the local hospital, had some familiarity with electronics.

Metcalf was quite intrigued with what he saw in Burlington and decided that this could be a natural means of expansion for a radio broadcaster into the new and probably competitive medium of television. After returning to Guelph he corresponded with Dr. Abajian and his brother Hank who, as an electronics engineer, had decided that CATV was the coming thing and was proposing to get into the business of building the amplifiers used on the Burlington system.

By December 1952 Metcalf had collected from several interested friends and associates the initial financing required and incorporated Neighbourhood Television Ltd. for the purpose of building a CATV system in Guelph. In the meantime, Jake Milligan had found a piece of land on the edge of Guelph where the only two available stations — WBEN-TV, Buffalo on channel 2, and CBLT, the new CBC station in Toronto on channel 6 — could be adequately received. He also checked into equipment suppliers and contacted Jerrold through Lou Harris of Atlas Radio, which was its exclusive distributor in Canada at that time.

Construction started early in 1953, using Jerrold strip amplifiers carrying the two available stations on channels 2 and 6 as received, and had progressed sufficiently for a sales campaign to commence with a public demonstration in a local school on the occasion of the coronation of Queen Elizabeth II on June 2, 1953. By the middle of 1954, after the first year of operation, Neighbourhood Television had connected 350 subscribers and had reached 1,000 by late 1955.

In building the system, it was necessary to bury about one-and-a-half miles of cable from the headend into town because of difficulties in working out an agreement with the Guelph Public Utilities Commission for the use of their poles. This delay was largely due to the fact that there was no legal authority in those days for cities to make agreements with cable companies, and it was necessary for the Ontario Legislature to pass an act authorizing a municipality to enact a bylaw for this purpose.

Early in 1956, when the system in Guelph was clearly a successful operation, Metcalf was approached by the Town Council of Huntsville, Ontario. In September 1955 CKVR-TV had commenced broadcast operation in Barrie on channel 3, providing the CBC network service from Toronto. Although Huntsville, located about seventy miles north of Barrie, was within acceptable reception distance of the transmitter and could, in theory, receive the signals, the town is situated in a river valley and no domestic reception was possible by any of the residents in the town.

An enterprising dealer erected an antenna on top of Reservoir Hill south of the town, on which he was able to receive the signals from CKVR-TV on channel 3 and then rebroadcast them using a low power transmitter on another channel. He and other dealers then proceeded to sell TV sets to the residents who could receive the signals from the local transmitter using simple rabbit-ear antennas. Unfortunately, he had not bothered to obtain a licence for this transmitter from the federal Department of Transport, which was the radio licencing authority at the time, and after selling something like one hundred sets the DOT closed him down. This action of course resulted in a considerable volume of complaints from the residents, especially those who had bought sets expecting local reception.

The town council had heard about the cable system in Guelph and invited Fred Metcalf and Jake Milligan to visit Huntsville to advise them on what would be required to build a system there. When cost estimates for a cable system were submitted to council, support for the idea quickly faded since no one, much less the council, was prepared to put up the money for what was considered a pretty risky venture. Finally Metcalf was asked if he would build a system. He agreed provided the council would permit the use of municipally owned poles at $1 per year each, and provide an area on top of Reservoir Hill which they owned to accomodate the headend.

Having agreed to build the system, Metcalf then needed some financial assistance. He approached Rediffusion in Montreal with the suggestion that they might supply, on a deferred payment basis, the strand hardware and aluminum cable. Rediffusion agreed, and as a result the system in Huntsville was the first CATV system in North America outside of Montreal to use solid aluminum-sheathed cable.

It was in Huntsville that the infant CATV industry had its first major confrontation with Bell. The system was constructed by agreement with the town council on municipal poles owned by the Public Utilities Commission, many of which were jointly used by Bell. Soon after construction started, Bell ordered the company to remove their cable from any of the poles which Bell was using. They had that right under the joint use agreement with the PUC, which had been written many years before and was a standard form of agreement with most public utilities.

The problem was brought to Huntsville town council and the mayor told the president of Bell that either the telephone company would agree to an amendment permitting the cable company to use the poles, or the town would order Bell off all municipally owned poles. The mayor offered Bell the alternative to build the system, but Bell declined and would not agree to permit joint use by CATV. The mayor then gave the telephone company six weeks to do one or the other. Finally Bell backed down and agreed to an amendment to the joint use agreement with the municipality permitting access by the cable company. This was the precedent for future joint use agreements between Bell and municipal public utilities allowing cable TV companies similar third-party rights on the poles. This change of heart resulted from the fact that the town really wanted the television service and were prepared to go to any lengths to facilitate it.

Much of the early pioneering work on receiving television signals and distributing them by cable took place in Ontario because, prior to the commencement of TV broadcasting in Canada in September 1952, most of the U.S. stations then on the air near to Canada were located in states adjacent to the Ontario border. However, all the early work was not confined to Ontario as evidenced by the story of Ed Polanski, who was the first pioneer in the province of Alberta.

When Ed Polanski graduated from high school in 1952, he was living in Thorhild, Alberta, a small town about forty-three miles north of Edmonton and 350 miles north of the U.S. border. While still in school he had been actively involved in amateur electronics because his father ran a hardware store in town and often received battery-operated radios which were in need of repair. This inspired Ed to consider electronics as a field he would like to work in and, reading about the expansion of television in the United States and plans to develop it in Canada, he decided to take a course in electronics to train for this future. He was unable to find any course in Alberta so in June 1952 he enrolled in the Radio College of Canada, located in Toronto. Television was included in the course curriculum, although the only television transmitter then receivable in Toronto was WBEN-TV in Buffalo. However, before he finished the course CBLT, the new CBC station in Toronto, began broadcasting.

Ed's intention was to get into the new field of television in the area of home receiver servicing. His objective was to set up a radio-television and electrical appliance sales and service store along the lines of the work he had been doing as a sideline in his father's store. After a year in Toronto studying electronics and television he returned to Thorhild in late 1953, and did in fact get into the service business alongside his father's hardware store, although initially this had to be confined to the kind of work he had done before — radio servicing.

While at the Radio College in Toronto, Ed Polanski had built his own television receiver which he brought back to Thorhild with him, only to discover that there was no television service available in Alberta even from the United States. Consequently, he started receiving and logging distant TV stations, not for their entertainment value because they were too sporadic for that, but rather for the satisfaction of achieving reception, however inadequate, over long distances. During the subsequent year-and-a-half he logged some forty-six stations from all over the United States, the furthest being in Amarillo, Texas, 1,400 miles from Thorhild. For this purpose he had a variety of antennas mounted on a ninety-foot tower alongside his store, since this part of Alberta is flat prairie country with no convenient hills nearby where he could get any height advantage.

At this time nobody in Thorhild or the surrounding country had ever seen a television picture or even a television set before, and farmers used to come in and sit in Ed's appliance store for hours just for the chance to see some freak reception. There were some stations, such as Great Falls, Montana, on channel 2, whose reception, although intermittent, was almost predictable on a daily basis, at least in the summer when weather conditions were more favourable for long distance reception.

In October 1954 CFRN-TV Edmonton started broadcasting, and Ed soon discovered that reception of the station on a domestic antenna was inadequate for regular viewing. Indeed the results even on his ninety-foot tower left much to be desired and were not helping the sale of sets in the store. So he moved his antennas to another site which happened to be at his father's home about a block away from the store, where he had found that reception was much better. He connected the antennas to the receivers in his store using a copper-pair cable, but found, as others had before him, that this lost too much of the signal and he replaced it with RG11 coaxial cable. Unknowingly, this was Polanski's first step into CATV, although his only intention at that time was to improve the quality of the reception in the store for the purpose of selling TV sets, just as John Walson and others had done in Pennsylvania six years earlier.

At this point neighbours and customers who had bought sets from him, and also antennas for receiving CFRN, were interested in getting the improved reception that Ed was able to demonstrate in the store. Having discovered that coaxial cable was the most satisfactory delivery agent available, he started to extend the cable from his antenna to willing customers who paid him $3.50 per month for this service with no installation charge. He had only three subscribers within the first year and even by 1958 still had a total of only seven. This extremely slow growth was not for lack of interest or demand, but was entirely due to the fact that he could not obtain access to the utility poles to install his cable; expansion was therefore limited literally to those he could serve by cabling from house to house. Again a repeat of Jarmain's early experience with the local utility in London, Ontario.

Even when pole access became available several years later Polanski did not extend the system in Thorhild because, as it turned out, the poor reception of CFRN was not due to the topography or distance from the transmitter, but rather to severe interference from the power lines which was experienced only in certain parts of town. In other areas reception was reasonably satisfactory, and Ed decided that the limited potential was not sufficient to justify the capital required to build a cable system.

In 1959 Polanski visited Athabasca on a curling bonspiel. He found there was an interest in the growing medium of television but no reception available from any station. Athabasca, about forty miles north of Thorhild and eighty miles from Edmonton, is located on river flats with hills some four hundred feet high to the south. These not only shielded the town from any TV reception from Edmonton, but also made radio reception extremely poor. In fact the town was almost isolated from a communications point of view, having only a weekly local newspaper and no daily paper from outside. Having discovered this market, and being interested in the potential of CATV, Polanski started to build a cable system in Athabasca in 1960.

By this time a number of CATV systems had been built in Canada, mainly in Ontario and Quebec, and a hardware technology was beginning to develop so there were several manufacturers producing amplifiers and associated equipment. Furthermore, coaxial cable had been developed using foam rather than a solid dielectric, providing lower losses than the cables used up to that time. As a result, Polanski was able to build a system in Athabasca which was technically superior to his first system in Thorhild.

The amplifiers, developed and manufactured by Benco, a Canadian company, were the first to use transistors instead of tubes and were the pioneers of the solid-state equipment now used exclusively. There were eight miles of network in the Athabasca system, with some fifty amplifiers, and the system served 250 to 300 subscribers paying $165 for installation and $5.50 per month for a single channel, CFRN Edmonton, no others being receivable even on the headend antennas. A second station, CBXT in Edmonton, was added in October 1961, and other TV stations, together with FM radio, were added as they became available.

In 1970 Polanski acquired the licence to build a system in Edmonton, which by 1977 had six hundred miles of cable with 36-channel capability, passed 80,000 homes, and served 60,000 subscribers. This indeed is a capsule example of the development of CATV over twenty-three years from 1954 to 1977, because Polanski's system in Edmonton was a far cry from his 300-subscriber, two-channel system in Athabasca in 1960, just as that was a far cry from his seven-subscriber, single-channel system in Thorhild in 1958.

In this capsule history of the early development of CATV in Canada we have been moving forward in time and westward in direction, starting with the first Canadian cable system in Montreal, Quebec, in 1951, through early developments in Ontario in 1952 and '53, and into Alberta in 1954. Moving further west we come to British Columbia where the pioneering spirit was just as strong.

The lower mainland of British Columbia, centred around Vancouver, had two things in common with southern Ontario back in 1952 — no television service of its own, and a station broadcasting television programs over the border in the United States far enough away, but not so far that it was impossible to receive its signal in the area. At that time the only television transmitter on the air in the U.S. Pacific

Northwest was KRSC-TV in Seattle, Washington, the station Ed Parsons in Astoria, Oregon, received from across the mountains on Thanksgiving Day in 1948.

Vancouver is about the same distance north of Seattle as Astoria is to the south — some 130 miles but, unlike Astoria, there are no high mountain peaks between them which could diffract the signals and give rise to weak but relatively fade-free reception. Instead the path is along the coastal strip and although a weak signal could be received in parts of the B.C. lower mainland, it required special antennas and a favourable location, and even then would be subject to considerable fading because of the distance.

In 1952 a company in Vancouver, Fred Welsh & Son Ltd., was involved in plumbing, heating, electrical contracting, and other related services concerned with the building and construction industry. The son, Syd Welsh, decided about this time to branch out into the sale and servicing of electrical appliances, including radio and television sets, and he incorporated a new company, Fred Welsh Home Appliances.

Syd Welsh at this time lived in a mountain-side home in affluent West Vancouver. This area lies above English Bay facing south towards the U.S. border and was therefore one of the more favoured areas for reception of television signals from Seattle. Even so, when some of his neighbours started buying TV sets it was necessary to use quite sophisticated antennas to receive pictures with any reasonable degree of quality and reliability. Syd realized there might be considerable business opportunities in the antenna field, and that in any case he would have to be in that business if he was going to sell many television sets. Again the same circumstances existed that motivated most of the other pioneers — facilitating the sale of television receivers where domestic reception was difficult or impossible.

Soon after the company was started it installed a master antenna system on a new luxury apartment building. It was not long before it was concentrating on the business of installing master antenna systems as well as the sophisticated domestic antennas needed to receive the Seattle station, in addition to selling receivers. In fact the direction in which the company's business was going was so specific that the name was soon changed from Fred Welsh Home Appliances to Fred Welsh Antenna Systems Ltd.

Very early in their television experience in Vancouver the company realized the potential inherent in cable distribution from a favourably located antenna. This was amply demonstrated by the master antenna systems it had installed. These were all on high-rise buildings where their big advantage was the location of the antennas on the roofs where they could "see over" higher ground or other obstructions blocking the signals from their lower neighbours. There were some areas in Vancouver where signals could be received and others where even the best antenna was practically useless. There is a ridge of high ground between Howe Sound on the north and the Fraser River on the south, and reception of the Seattle station was generally practical, although often unreliable, on the south slope facing towards Seattle, but virtually impossible on the north slope where it was shadowed by the ridge.

About this time in 1952, George Chandler, who owned radio station CJOR in Vancouver, had also realized the potential of CATV and had started a small cable company which he called Tru Vu Television. Chandler had succeeded in negotiating a contract with B.C. Telephone for the right to attach his cables to the telephone poles owned by the utility in Vancouver. This was one of the first, if not the first, such pole contract in Canada. Not only did he have this attachment right, but telephone crews installed the cable, amplifier housings, and passive equipment for him, and ran the subscribers' drops to the homes, all on a contract basis, so that he retained the ownership.

Chandler's headend was located at 59th and Heather Streets on the south slope in one of the locations where reception of KRSC-TV Seattle on channel 5 was practical. By July 1953 some twenty-three miles of cable had been installed with twelve or thirteen trunk amplifiers of the SKL distributed type, similar to those being used by Jarmain in London, Ontario. These were probably the first CATV amplifiers accommodated in pole-mounted cast aluminum housings. They were unusual because in those early days the amplifiers were invariably housed in heavy sheet steel boxes with a hinged and flanged lid on top, and were mounted horizontally across two cross-arms on the pole. The subscriber taps used on the system were also of SKL design, each tap consisting of two-way splitters cascaded to provide outputs to eight subscriber drops. Fortunately, amplifier output levels were much higher than on those used today. All the connections were soldered in, inconvenient but less conducive to radiation problems than many of the connections used in subsequent system construction.

At this time the telephone company was not prepared to give pole attachment rights to a second company in the same area, so Fred Welsh Antenna Systems could not develop the potential of cable distribution in its home city beyond the master antenna systems on individual buildings. Nevertheless, the company knew that there were many other communities in British Columbia which had no means of receiving television but were probably within reach of mountain locations at which the signals from the Seattle station could be received. So, using the antenna experience it had acquired in Vancouver, it set out to try to realize the potential of cable distribution in other areas.

Garth Pither, as the technical partner in the company, spent much time surveying potential antenna sites for useable signals, and when found he figured out ways of getting those signals by cable to a nearby community, and then designed and installed the distribution system to make them available to the homes. In this way, while still unable to develop cable distribution as it would have wished to in Vancouver, Fred Welsh Antenna Systems did install systems in Youbou, Courtney, Comox, Squamish, Trail, Kaslo, Penticton, Rossland, Kimberley, and a number of other small communities in British Columbia.

Although this proved to be a profitable business, the company was building these systems for others on a contract basis and Syd Welsh could see that operating cable systems himself could be potentially more profitable than simply installing them for others. Thus, by 1958 the company decided it had to be in that business, and in Vancouver where there was a huge undeveloped potential. It was successful in obtaining pole attachment rights from B.C. Telephone, and incorporated Pacific Cablevision Ltd. During the next four years, from 1958 to 1962, it expanded its cable system in Vancouver as rapidly as limited financial resources would allow, and this question of financing proved to be a major limiting factor.

There were very few financial institutions with faith in this fledgling industry in those early years. Banks, insurance companies, trust companies, and other institutional investors could not see the future potential in CATV and simply were not interested. It was therefore necessary to rely on the installation charges paid by subscribers as they were connected for the capital required, and on private entrepreneurial investors, and Syd Welsh had to seek additional partners as a source of capital. As a result expansion took place in a series of pocket areas, since each new area built was financed with the help of new partners in the area concerned.

It was not until 1962 that Welsh was able to obtain substantial financing from Laurentide Finance, which enabled him to buy out George Chandler and consolidate the Tru Vu Television system with that of Pacific Cablevision into a single system serving the Vancouver area under one company, Vancouver Cablevision Ltd. Tru Vu by this time had expanded to a system with seventy miles of cable serving approximately 7,000 subscribers, and the service it provided had been expanded to a total of seven channels from the original single channel. While this consolidation improved the cash flow available for further expansion, it did not relieve the constant need for more capital as there was still a very large undeveloped potential, and Welsh knew that future expansion would be in direct proportion to the amount of capital that could be raised.

A chance encounter in 1963 between Bud Shepard, one of Welsh's original partners, and Harvey Struthers of Columbia Broadcasting System at a convention in Seattle brought the final solution to this problem. CBS, a major television broadcasting network in the United States, was farsighted enough to see the potential in cable television, or perhaps could see the potential competition in television program distribution with their affiliated stations, and was interested in becoming part of it. However, CBS was apprehensive about the possible negative reaction from their network affiliates if they were to follow this course in the United States. Participating in cable development in Canada was therefore the perfect answer. An agreement was negotiated whereby CBS acquired 75 percent of the capital stock of the Vancouver company, Canadian Wirevision Ltd., with Welsh and his associates remaining as top management and, most important to the Canadian partners, CBS putting up all the financing required for further expansion.

In 1969 a federal government order-in-council, requiring all cable licencees to be at least 80 percent Canadian-owned, forced CBS to divest themselves of all but 18.6 percent of their equity in the Vancouver-based operation in which by then they had invested some $18 million. However, this timely infusion of capital had enabled Vancouver Cablevision to expand their cable TV interests to the point where, by 1971, they were able to go public with the formation of Premier Cablevision Ltd. and become one of the larger multiple system owners in Canada.

CHAPTER 5

The development of pay-TV

Community Antenna Television or CATV was an appropriate description because this was exactly what it was — an antenna system shared by the community. Its sole purpose was to make available to each home in the community the TV signals which could be received on this common antenna and which were not as easily receivable on a domestic antenna located at each home. The signals carried by the system were those received on the antenna and no others. Indeed it was sufficient to be able to view broadcast television programs which were not capable of being received on the individual homes, and there was no requirement to use the cable for any other purpose.

As with any other newly developed system, things do not stand still for long, and familiarity breeds a desire for more. This has always been the case with an entertainment service such as television where the broadcaster has the choice as to the programs and the times at which they will be shown. When broadcasting first introduced this new medium to a community the novelty alone was sufficient to ensure acceptance of almost anything the broadcaster offered. But inevitably this condition did not last for long, and the viewers felt the need to be able to excercise some choice for themselves. This was the motivation for the early CATV systems — to reach out with increasingly sophisticated antennas for stations to add to the service in order to provide such a choice.

Limited off-air reception was not the only factor inhibiting viewer choice. From the earliest days television programming, at least in North America, has been primarily financed by advertising. The nature of this financial support has meant that the program material presented is largely determined by the advertiser — an application of the old adage that he who pays the piper calls the tune. It has long been recognized that one way to give program choice to the viewer is to transfer the financial support to that end of the system, to have viewers pay for the programs they want to watch when they want to watch them. This is known as "pay-television." This form of financial support for TV programming provides a vehicle for escaping the strait jacket of commercial broadcasting tied to mass audience appeal and, at least potentially, making available program material more suited to selective minority audiences rather than a mass audience.

To most people pay-TV is a recent development. Indeed as a practical commercial service it is, since it owes its present growth to computers, micro-electronics, and satellite communications, technologies which did not exist commercially during CATV's early growth. However, the concept of pay-TV as a service of consumer choice and an alternative to advertiser-financed programming was conceived in the early days of television broadcasting and was the subject of a number of experiments and practical field trials during the years that television was expanding as a commercial public service.

It is worthwhile to review the history of these experiments because, although none of them developed into full-fledged commercial systems large enough to be economically viable, they illustrate the fact that pay-TV was an idea waiting for the development of a suitable technology, and that the service itself was, and still is, the subject of conflicting philosophies.

Zenith Radio Corp., a major manufacturer of television receivers and electrical products, announced the earliest development of a pay-TV system in 1947. At that time there were no more than fifteen television transmitters on the air in the United States, and none at all in Canada. This suggests that, even in the very early days of television broadcasting when the medium was still a novelty, there were people who had serious doubts about the acceptability of a public entertainment service wholly supported by commercial advertising, and they were prepared to invest in the development of an alternative system. It should be noted that Zenith at this time did not consider the application of pay-TV to wired systems, which were still very few in number and far from being technically sophisticated. The company's developments were therefore essentially with broadcast TV in mind.

Zenith called this early development Phonevision. As the name implies it required the use of a telephone whereby the subscriber indicated to a central control the desire to receive a scheduled program. The broadcast signal was scrambled and could not be unscrambled until the subscriber telephoned the order. A specially coded signal was then applied to a decoder connected to the television receiver to unscramble the picture.

The first field trial of Phonevision occurred in Chicago in 1951 with some three hundred viewers to whom decoders had been supplied. Zenith's experience in this trial convinced them that any viable system of broadcast pay-TV required scrambling of the sound and a more sophisticated picture scrambling system. Since the broadcast signal was receivable by every set in the coverage area of the station, representing about 34 percent of the households, security of the signal was of great importance right from the start.

During the trial it soon became apparent that the commercial-free material was attracting a large audience without decoders who could not follow the scrambled picture but who were enjoying the unscrambled sound as they would a radio broadcast. Furthermore, many found that, even though the picture jitter caused by scrambling was annoying, they were able to derive enough information from the scrambled picture to make the program attractive so long as they could follow the unscrambled sound. When Zenith scrambled the sound and introduced a more complicated form of picture scrambling, use of the service as an entertainment medium was impractical without a decoder. Thus, the need for excellent security against pirating of the programs was proved almost in the first trials.

The programming on Phonevision consisted exclusively of movies available after their theatre run and generally at least two years old. In those days movies were not available to commercial TV, and surveys indicated that only 20 percent of Phonevision subscribers had previously seen them in a theatre. So they had a strong appeal, particularly as there was not the wide choice of program material available on commercial television that there is today. A different movie was offered every evening at $1 per movie, and this was simply added to the subscriber's telephone bill. The service was clearly popular, but impulse buying, coupled with the novelty of the service, inflated the phone bills beyond a level that the subscribers were prepared to accept, and the experiment was discontinued, having to all intents strangled on its own success. Nevertheless, Zenith was encouraged by the results of this first trial and only deferred further commercial trials while continuing development of the hardware.

Then in June 1962 Zenith initiated the first commercial broadcast pay-TV service in the world in Hartford, Connecticut as a joint venture with RKO General on their station WHCT. A scrambled signal was broadcast and each subscriber had a decoder attached to the TV set which could unscramble the signals. In order to watch the program, the subscriber adjusted five dials on the decoder to coincide with an "air code" transmitted with the program signal, and then pressed a button which marked a tape inside the decoder identifying the program watched. The subscriber sent the tape in monthly together with payment for the program charges and the decoder rental.

Subscribers paid $10 for the installation of a decoder plus seventy-five cents per week service charge, and from fifty cents to $3 per program selected for viewing. Approximately forty-five hours of pay-TV programming were broadcast each week, some 87 percent of which consisted of movies, and the balance live sports and special entertainment. Initially the service proved very popular. During the first two years the demand exceeded the supply of decoders, while the program revenue per subscriber averaged $5 per month. However, the subscribers peaked at 7,500, or 3 percent of the TV households in the station's coverage area and, since this was not sufficient to break even, the project was discontinued in 1969. Later Zenith claimed that reduced hardware costs made possible by newer technology might have made the system viable at that penetration level. Another factor which hurt the Hartford operation was that the system had been developed for broadcasting in black and white and could not be readily adapted to colour, and subscribers with colour sets would not pay for movies in black and white when they could watch programs in colour on commercial TV for free.

Another early pay-TV development was the Skiatron system, which the FCC authorized for experimental broadcast in New York in 1950, although this authorization was never implemented. This was very similar in its basic concept to the Zenith system, except that it did not use the telephone for upstream communication and billing. Like the Zenith system it scrambled both picture and sound, with the unscrambling effected in a decoder associated with the subscriber's receiver, but in this case the descrambling was under the control of a punched card inserted into the decoder.

The Skiatron decoder accepted an IBM punched card with holes at coded locations, through which contacts in the mechanism could be activated. Direct use of the holes was avoided to restrict duplication of the card and possible counterfeiting of the codes. Instead, the punched holes were used in conjunction with printed circuitry superimposed on the card in such a way that detection of the code was impossible until the card had actually been inserted in the decoder and used. Until this time many of the contacts on the card were electrically shorted together and the code itself scrambled. With the card inserted and the decoding button pressed a punch was actuated, breaking these shorts and presenting the correctly coded combination to the mechanism.

Each card contained thirty individual codes on its surface so that a single card could be used for, but not limited to, one program per day for thirty days, thus permitting the same card to be used for a month. In addition to providing decoding information the card also gave details of the programs offered, together with prices and times of exhibition, and served as a billing vehicle. Initially a subscriber received cards for two months of service. The first month's card was mailed to the company with the payment for the service and program charges. Before the second month was completed, the subscriber would receive a card for the following month to which was attached a stub and the total cost. The stub would then be detached from the card and the card inserted into the decoder at the beginning of the month to which it applied.

Skiatron did not get around to an actual field trial of the system in conjunction with a television broadcaster until 1964. It was introduced in San Francisco and Los Angeles simultaneously with a much more ambitious project by Subscription Television Inc. (STV), which had started in 1963, involving the installation of dedicated cable for pay-TV distribution in high-income areas of these two cities.

The programming offered on the STV system was on a pay-per-program basis with strong emphasis on sports. In fact baseball games accounted for 44 percent of the schedule, while movies accounted for only 33 percent, and educational type programs made up the remaining 23 percent. Baseball predominated because two of the STV shareholders were Walter O'Malley and Horace Stoneham, owners respectively of the Los Angeles Dodgers and the San Francisco Giants baseball teams. Both teams had moved to the west coast in 1957, and since then their owners had refused to negotiate rights to their games with commercial television because of their belief in the future of pay-TV and their expectation that this could be exploited to their advantage.

It was not long before STV had 6,500 subscribers in Los Angeles, and 2,000 in San Francisco, with subscriber revenues averaging $15 per month. This was a high figure for those days and much higher than had been experienced in the earlier pay-TV experiments, but the system's promise of success sounded its own death knell. It aroused great concern in the movie theatre world, already suffering heavily from the competitive inroads of commercial television with its increasing use of movies. The National Association of Theatre Owners, also based in California where this competition was most apparent, started a well-financed campaign resulting in the inclusion of an anti-pay-TV referendum on a state ballot in November 1964 which was passed by a huge majority.

The 1964 referendum resulted in a ban on pay-TV in California, and Subscription Television Inc. had to close down, losing over $10 million. The California Supreme Court later overturned the ban as being contrary to the First Amendment guaranteeing free speech, but by this time it was too late for STV and the project was never revived.

Initially pay-TV developed as an alternative to advertising as a method of financing a broadcast television service. Since broadcast programs are available to all receivers within reach of the transmitting station, some form of security had to be an essential part of the system. Thus, signal scrambling and controlled decoding were an inevitable feature of each of these systems to ensure that the programs were only available to those who paid for them. Although there were companies with the foresight to see that the day would come when the public, to whom at that time television was still a novelty, would begin to tire of the banalities of commercial television and be prepared to accept an alternative method of financing and presentation, there were some regulatory roadblocks in the way.

Since the Communications Act of 1934, the Federal Communications Commission (FCC) has closely regulated broadcasting in the United States. The FCC was very reticent to approve any form of pay-TV applied to the broadcasting system because of its politically controversial nature and the existence of a strong broadcasting lobby. The Commission approved experiments by Skiatron and Zenith on a limited basis in 1950 and '51, but did not issue any more approvals, even for limited experimental purposes, until 1959. Thus, other companies turned their attention to the possible use of cable distribution for pay-TV programs. At that time cable was not subject to the regulatory control of the FCC. Furthermore, by its very nature and its limited availability, cable appeared to require less concentration on the security aspects of signal scrambling and controlled decoding.

Indeed it is interesting that while entrepreneurs in the east were experimenting with the use of cable distribution as a means of making commercial broadcast television signals available to homes which they could not otherwise reach, others in the west were developing systems to distribute TV programs by cable to subscribers prepared to finance them on a user-pay basis. The first of these was International Telemeter, a company based in Los Angeles and 50 percent owned at that time by Paramount Pictures Corp.

International Telemeter had been developing equipment for pay-TV distribution by cable since 1949, and the system was based on the simplest form of payment in the home — a coin box. It was this basic principle that first attracted the attention of Paramount, a major producer and distributor of motion picture entertainment for almost half a century. Based on this long experience Paramount was firmly of the opinion that mass entertainment could only be successful when paid for on the spot in cash. Credit and billing had never been their way of successfully selling entertainment. The public prefers to pay for amusement as the mood strikes since it is essentially an impulse purchase. Paramount claimed this was the verdict of the history of theatre, movies, and sports, and it was convinced that this offered the only practical approach to the marketing of pay-TV that was consistent with its experience in serving a mass market.

The design of the Telemeter system was therefore based on a coin box which was part of an attachment associated with the subscriber's receiver and could accept several denominations of coins. Since Paramount was also convinced that any system of pay-TV required an adequate means of identifying the programs purchased in order that the revenue may be properly divided among the various interested parties, the attachment automatically identified and recorded the program selections and the amounts of money deposited in payment. This recording on tape was collected periodically with the money in the cash drawer and formed an audit of the cash collection and the basis of payments to the program producers.

Another feature of the Telemeter system was the provision of a "marquee" and a "barker." These were colloquial terms derived from theatre exhibition. The marquee gave a visual announcement of the programs being shown on each channel, program price, and times of showing, together with any other pertinent information, while the barker duplicated this information orally. This information was available continuously to all subscribers. When a program had been selected for viewing and the required coins deposited, the marquee and barker would be removed and replaced by the selected program.

The Telemeter system was designed specifically for use on cable and was neither applicable to, nor intended for use with, the broadcast medium. It did not scramble the transmitted signals, but instead ensured the required security by transmitting at channel frequencies which were outside the range of the TV broadcast frequency allocations and could not therefore be received on standard sets connected to the cable. A Telemeter subscriber was supplied with an attachment to be connected to the TV set which incorporated, in addition to the coin collection and program recording mechanism, a frequency converter capable of changing the special pay-TV channels to a standard VHF channel to which the set could be tuned, generally channel 3 or 4 as in present-day converters. If the subscriber did not have the exact change, the coin box was able to collect coins to a larger amount than required for payment for a selected program and to credit the excess against the purchase of a future program.

By 1953 Telemeter had completed their hardware development to the point where they wanted a commercial field trial, and for this purpose they built a cable system in Palm Springs, California. Palm Springs is a wealthy resort community in the desert some one hundred miles from Los Angeles, and surrounded by mountains high enough to cause problems with the reception of television from the nearest transmitters located on Mount Wilson outside Los Angeles. The cable system distributed the Los Angeles stations received off-air at a suitable antenna site and subscribers were charged an installation fee of $150 plus $5 per month.

Pay-TV was then added to the system and offered as an optional service. Subscribers paid $21.75 for the Telemeter attachment, plus a charge of up to $2 per program selected, depending on the program, with a minimum charge of $3 per month. Despite the relatively high cost 512 subscribers were connected in the early days, and seventy-one of these were equipped with Telemeter attachments when the pay service was inaugurated. By February 1954 subscribers were signing up for the Telemeter service at the rate of twenty per week.

However, first run movies were the staple programming and these proved so popular that the local theatre owners began pressuring the film distributors not to supply films to Telemeter. After a drive-in theatre owner threatened a lawsuit all the distributors, except Paramount, bowed to this pressure and refused to supply product. Paramount of course continued to supply the movies to which they had the distribution rights but these were not sufficient in number to support the service. In April 1954 the operation was closed down, with 170 paying subscribers still spending an average of $8.70 per month for the pay programs.

Telemeter claimed that the trial had been a financial success and told the FCC that "this test provided initial evidence that, given special entertainment and sports not otherwise available on commercial television, the public would respond and pay for such home entertainment at a rate which would justify the operation from an economic standpoint." However, the results were inconclusive because domestic reception of the commercial television stations from Los Angeles was of poor quality, making the CATV service itself attractive, and this, coupled with the high average income of the Palm Springs residents, did not constitute a valid test of public acceptance of the pay-TV concept.

The next commercial trial of pay-TV by cable occurred in Bartlesville, Oklahoma, in 1957. This experiment was planned by Video Independent Theatres, a company owning some 150 movie theatres in Oklahoma and Texas, including two conventional theatres and two drive-in theatres in Bartlesville. The company also had interests in six CATV systems, a television station in Oklahoma City, and a radio station in Sante Fe, New Mexico.

Video Independent Theatres decided in 1956 to build a dedicated cable system in Bartlesville, intended virtually as an extension of the movie theatres into the homes. The programming of the system would follow closely along the lines of normal movie theatre presentation; it would present first run and similar feature movies on a continuous play basis, but exibited in the subscribers' own homes instead of in the theatre — in effect an extension of the regular theatre.

Nevertheless, there was one basic difference in the method of operation which eventually proved to be the main undoing of the experiment. In spite of Video Independent's long experience in the entertainment field, and contrary to the principle adopted earlier by Telemeter based on Paramount's parallel experience in theatre entertainment, it was opposed to a coin box system of cash collection as it felt that this method would be too costly to operate, and would be resented by the viewing public.

This opinion was based largely on the almost complete absence at that time in the United States of any system of coin collection by the public utilities in favour of billing, making this a familiar and universally accepted system of charging for services delivered to the home. It was felt, with some justification, that requiring cash payments in advance, rather than billing in arrears, would introduce another element of uncertainty in an experiment designed to test the acceptability of the movies-in-the-home or pay-TV concept. Subscribers to the Bartlesville system would be billed at a flat rate of $9.50 per month. For this they received a total of twenty-six movies each month, the charge being independent of the number of movies watched.

Bartlesville was carefully selected as the location for this experiment for several reasons. The town had a population in 1956 of 28,000 with 5,800 homes, of which 5,200 had television sets. The average family income was considerably above the U.S. national average. The average monthly expenditure on movie theatre attendance in the town was $2.50 per family. In addition, off-air television programs were available from three transmitters in Tulsa, fifty miles away, which were well received on domestic antennas, providing the three U.S. networks. The selection of this combination of circumstances was intentional as it was felt that if a cable movie operation could succeed under these conditions then it was likely to succeed in almost any community.

Jerrold Electronics devised the engineering design and installed the system. Technically it consisted of a standard CATV system so far as distribution was concerned, but was fed from closed-circuit transmitters and film projection equipment at a central studio instead of being fed from off-air receiving equipment at an antenna site. The three Tulsa stations available off-air were still received on each subscriber's own antenna. A switch was installed on the back of the TV set to provide the choice between free television over the air or tele-movies over the cable. In other words, the system was designed specifically as a means of distributing movies to the home and was not intended to provide a CATV service.

Two tele-movie programs were distributed simultaneously on channels 3 and 5, with facilities provided to add a third later on channel 6. The programs were not scrambled since it was considered that the security offered was the same as that of a normal CATV system. This was probably valid since subscribers were only connected to the system to obtain the movies and were not using it for reception of the broadcast television programs with the movies as an optional extra service.

Studio equipment comprised two separate film chains, each using two 35 mm projectors with industrial type vidicon cameras specially modified to handle Cinemascope and Vistavision wide screen films in addition to standard stock. A third film chain was also installed ready for use on the third channel when it was activated, but this was never brought into use. It was necessary to use 35 mm projectors because all film prints for distribution to theatres were on this size stock, and if 16 mm prints were made at all it was not normally until at least sixty days after theatrical release. Thus, 35 mm facilities had to be available if first-run theatrical features were to be shown.

First-run features were shown on one channel and reruns on the second on a continuous performance basis from about midday to 11 P.M. daily. On each channel the features were changed every two or three days, and in the case of first-run movies they were generally showing at one of the local theatres owned by the company, but not simultaneously.

The Bartlesville experiment began on October 1, 1957. By December some 550 subscribers had been connected, but it soon became clear that the rate of connection was not nearly as great as had been anticipated. By February 1958 steady cancellations had reduced the number of subscribers to three hundred. At this point the monthly charge was reduced from $9.50 to $4.95, and the movie programs were confined to one channel instead of two with continuous performances from 7 to 11 P.M., eliminating daytime operation. The second channel was then used to provide a time and weather service with Muzak background music on the sound from 7 A.M. to 11 P.M. In addition CATV service was added using the three Tulsa stations so that subscribers did not require their own antennas for off-air reception.

These changes increased the number of subscribers to about six hundred within the next month or so, but the field trial was finally suspended on June 6, 1958, although by that time the number had reached a peak of 765. The main reason for the suspension was the fact that the cable facilities, leased from Southwestern Bell Telephone, were too expensive and made the charge of $4.95 uneconomic for any subscriber total which could reasonably be expected in Bartlesville. The break-even point at this income level was some 2,000 subscribers, or 40 percent of the total television households, and 15 percent was the maximum penetration that had been achieved.

As a result of their experience with this system, Video Independent Theatres concluded that a flat monthly charge did not meet with public acceptance. The public clearly preferred to pay by the picture and not on a monthly "package" basis. There is a strong psychological element in this because, even at a monthly charge of $9.50, if a family watched a majority of the twenty-six movies offered, on a per movie basis this was very cheap; however, if a subscriber was unable or unwilling to watch many of them then the cost per movie increased to an apparently unacceptable level.

Even so, Video Independent Theatres was still strongly of the opinion that subscribers would react against any attempt to collect payment on a cash basis, for example by means of coin boxes as in the Telemeter system, and that they should be billed for this service in the same way as they were billed for telephone, gas, electricity, and CATV services. VIT had indeed come to the conclusion that the right answer was a per program metering system and was strengthened in that opinion by the difficulties it had experienced in determining the division of the revenue among the various distributors supplying the product. It also recognized that if future programming were to include sports and other special events some form of metering, and possibly some method of flexible pricing, would be necessary.

Based on this experience in Bartlesville, and the conclusions drawn from it by Video Independent Theatres, Jerrold Corp. developed a billing system which could be applied to pay-TV or cable movie operations so that the fixed monthly charge could be avoided and instead subscribers could be charged only for the programs they selected to watch. They called this the "PBPB," or "Program-by-Program Billing" system.

In retrospect, the PBPB system is of interest because it employed techniques which, with more advanced technology, came into use on cable systems more than thirty years later. In fact the system, in concept at least, was probably well ahead of its time. Instead of using a coin box and register tape in the subscriber's premises like the Telemeter system, or a punched card decoding mechanism like the Zenith, Skiatron, and other contemporary systems, it concentrated its control and billing functions at the sending end of the distribution system.

Each subscriber's unit included a crystal-controlled solid-state oscillator capable of returning a signal to the central control using a frequency between 500 and 2,500 KHz, which was discrete to that subscriber. The subscriber's program selector activated the oscillator and the presence of an output indicated the subscriber was watching that particular program. The units were continuously interrogated by a mechanical scanning receiver, which could cover 2,000 units in just over one minute. The presence of a signal on a specific frequency indicated that the subscriber to whom that frequency had been allocated was watching that program. This information was then recorded, either on paper or magnetic tape, by means of a binary code representing the subscriber's name and address. This in turn would be used for billing. Today a similar scanning and recording process would be undertaken electronically by a computer, and the same principle is being used in addressable taps and other interactive devices. So far as is known this system was never tried out except perhaps on a small-scale experimental basis, and certainly it was never applied to a commercially operated system.

In the meantime, International Telemeter had been quietly proceeding with further development of their pay-TV system in Los Angeles. Following the early termination of the Palm Springs test in 1954, the company had been looking for a more suitable location for an experimental system. The problem in Palm Springs arose when the supply of program material dried up due to the close association between Paramount Pictures, as a movie producer, and the theatre exhibitors. The problem was caused by the U.S. "consent decree" which forbade producers and distributors from owning theatres and thus providing a ready market for their own product to the detriment of competing exhibitors. This situation induced Paramount to look at the possibilities in Canada where no such embargo existed. Indeed, the reverse was the case because Paramount Pictures owned a controlling interest in Famous Players Canadian Corp., and Famous Players owned the largest chain of movie theatres in Canada and could therefore be counted on for cooperation so far as supply of product was concerned.

During the latter part of 1957, International Telemeter and Rediffusion Inc. discussed the possibility of a commercial trial of the Telemeter system on Rediffusion's cable network in Montreal, Canada. The Telemeter system distributed three video channels at frequencies between 10 and 40 MHz (below the broadcasting band) without scrambling, plus audio and control signals at around 8.5 MHz. The amplifiers were very similar in electrical design and physical configuration to those used on the Rediffusion system, which also distributed three channels in the same frequency band. Thus, the two systems had a substantial degree of technical compatiblity. For International Telemeter such an arrangement offered major advantages, not only in the availability of suitable distribution equipment and facilities already in place but also in the extensive experience which Rediffusion could offer in the planning and operation of a cable system.

In November 1957 Rediffusion Inc. and Paramount Pictures, the parent company of International Telemeter, verbally agreed to the cooperative venture. Paramount agreed to invest $2.5 million to purchase a 50 percent interest in Rediffusion Inc. Rediffusion in turn would provide the facilities for a large scale pay-TV trial on its Montreal system.

This verbal agreement was never formalized. While these discussions were taking place in New York and Montreal, Rediffusion's parent company in England had been approached by Skiatron which was equally anxious to undertake a commercial trial in what they considered to be their prime markets, Los Angeles and San Francisco. Skiatron's proposal to Rediffusion in London consisted of an offer to purchase a substantial interest in the Canadian company, while Rediffusion Inc. would provide technical services to Skiatron in the development of pay-TV systems on cable in Los Angeles and San Fransisco, and later in other locations in North America.

Rediffusion accepted the Skiatron proposal in preference to Paramount's. Organizational changes then proceeded in Rediffusion's Montreal office in preparation for the considerable amount of work which would be involved in planning for the first two major pay-TV systems. This included the assembly of training facilities for the personnel expected to be hired for design, construction, and operation of the systems. In spite of this flurry of activity extending over several months, nothing further developed because Skiatron failed to put up its share of the capital cost which was due within three months of completion of the agreement. Although Rediffusion extended the deadline for another three months, Skiatron's contribution was still not forthcoming by late in 1958 and the whole plan collapsed.

By the end of 1958, Rediffusion was left with no plan for participation in the further development of pay-TV on cable, and International Telemeter still did not have the test bed they had hoped for free of restrictions on the adequate supply of program material. With the failure to successfully negotiate the use of the existing cable system in Montreal, International Telemeter was convinced that the best chance for a successful field trial, which would not attract the opposition experienced in Palm Springs, still lay in Canada, and that the answer would be to build a distribution system for the purpose with the cooperation of their associated Canadian company, Famous Players.

To this end, in 1958 Telemeter, in cooperation with Famous Players, decided that a cable system covering some 5,000 homes for the specific purpose of delivering closed-circuit pay-TV programs would be built in London, Ontario. Famous Players chose London for several reasons. It was within convenient distance of the company's base in Toronto, was a reasonably affluent urban area, and at that time had only one local television station available with no other station within domestic reception distance to provide off-air competition for viewers. In fact these were the very circumstances that provided the inducement for Ed Jarmain's experiments in long distance reception and cable distribution in this area, and his ultimate decision, at about the same time in 1958, to build a commercial CATV system.

However, before construction of the proposed pay-TV system began in London, International Telemeter realized that from a market test point of view a successful pay-TV experiment under these circumstances could be very misleading, perhaps even meaningless. Critics could too easily suggest that its success was due to lack of competition, and there would be no guarantee that it could succeed in a more competitive environment. It was therefore decided to build the experimental system in an area with similar demographics in metropolitan Toronto. Off-air television competition would be present in the form of three U.S. network stations in Buffalo, N.Y., two Canadian stations, one in Toronto and one in Hamilton, with a third Canadian station expected to be on the air in Toronto within two years.

The area selected was in Etobicoke, a bedroom suburb of Toronto, immediately west of the Humber River and centred round Bloor Street and Royal York Road where the administration offices, studio, and program production facilities were to be located. Trans Canada Telemeter, a wholly owned subsidiary of Famous Players Canadian Corp. set up specially for the purpose, would own and operate the system. When this decision was made Famous Players had already made substantial commitments in London, particularly to Bell Canada for cabling, and, rather than cancel these, decided to cooperate with Ed Jarmain in the construction of a regular CATV system as described in Chapter 4.

Construction of the pay-TV system in Etobicoke started in 1959 and was completed ready for commencement of public service on February 26, 1960. Bell Canada installed the coaxial cable under a normal CATV-type partial system agreement in which Bell leases the cable to the operator, who installs and maintains his own amplifiers and subscriber drops. Trans Canada Telemeter installed the subscriber drops and also the amplifiers which were supplied by International Telemeter. The program distribution centre at Bloor Street and Royal York Road included a control room housing the transmitting, monitoring, and control equipment, a projection room with three 35 mm tele-cine film chains and two Ampex two-inch quad tape machines, and a well-equipped studio for live productions. Thirty-five mm projectors were used in the film chains because the films to be shown were regular theatrical releases as distributed to theatres, and at that time at least 16 mm prints were not available until some time after theatrical showing.

Within the first year of commercial operation the service proved to be very popular, some 5,600 subscribers being connected. Since there were approximately 14,000 homes in the wired area capable of being served, this represented a saturation level of around 40 percent. Apart from the special features of the pay-TV service this was a large cable TV system for those days, probably exceeded only by the Rediffusion system in Montreal. Certainly no other CATV system approaching this magnitude existed in North America at this time.

The basic program fare on the Telemeter system was movies made for theatre exhibition and, since Famous Players was one of the two leading distributors and exhibitors in Canada, there was no problem obtaining product as there had been with the earlier Telemeter experiment in Palm Springs. The movies shown were not first release, but rather first suburban release; in fact they were frequently playing day-and-date with the suburban theatres.

However, it had always been intended that this fare would be supplemented by live entertainment specially produced for the pay-TV service and sports events not available on broadcast television. This took somewhat longer to organize and the first non-movie program offered to subscribers was football from the Canadian National Exhibition (CNE) stadium in Toronto on August 3, 1960. This was followed on November 13 by an away hockey game of the Toronto Maple

Leafs from Boston, transmitted to the Telemeter distribution centre by microwave. In the meantime arrangements were under way for special live entertainment productions, but these took longer to organize and could not be staged that frequently because of the costs involved in relation to the very limited subscriber market at that early stage of development.

The first live show presented was *An Evening With Bob Newhart*, who has since become a popular and regular entertainer on network television. This was an intimate type of show in night club style, staged in the Telemeter studio before a live audience on January 5, 1961. Next was the Broadway musical *Showgirl* starring Carol Channing, which was videotaped on stage in New York and first presented to Telemeter subscribers on April 2, 1961.

It was nearly a year before the next live productions were staged and these were on a somewhat larger scale. In fact the Telemeter studio was not large enough and instead a studio was used at CFTO-TV, the new Toronto TV station which had just opened the previous January and was to become the flagship station of the CTV network. On February 23, 1962, a two-hour show was produced with Bill Dana and the McGuire Sisters which was transmitted live to the Telemeter subscribers and also taped for repeat showings, and on March 9 another live program was produced from the same studio, featuring the very popular British performers Gracie Fields and Stanley Holloway.

In the meantime negotiations had been successfully completed to telecast all the away games of the Toronto Maple Leafs. The home games played at Maple Leaf Gardens in Toronto had for some years been broadcast by the CBC as "Hockey Night in Canada." But these broadcasts had never been extended to the games played in New York, Boston, Chicago, and Detroit, the home bases of the four U.S. teams then in the National Hockey League.

At this time Famous Players was looking beyond the Telemeter pay-TV system, confined to a few thousand homes in Etobicoke, as the vehicle for exhibiting these games and was actively working to include movie theatre and arena audiences using large-screen TV projection. At their plant in Einhoven, Holland, Philips Electronics was producing the Eidophor video projector, capable of projecting a TV picture onto a screen twenty to thirty feet wide and fifteen to twenty-three feet high with excellent brightness and definition, and arrangements were made to obtain one for experimental and demonstration purposes.

In March 1962 a survey was conducted of Maple Leaf Gardens to determine the requirements for an Eidophor installation there. A couple of months later Famous Players' executives witnessed a very successful demonstration of the Eidophor projector operating under similar conditions in Madison Square Gardens, New York. Later that year the demonstration projector was installed on a temporary basis in Famous Players' College Theatre in Toronto and on October 28 carried a hockey game from Detroit live before a theatre audience, followed three days later by a game from Montreal. CFCF-TV handled the pick-up of the game in Montreal, and the games were transmitted direct to the theatre in Toronto by telephone company microwave.

Following these successful demonstrations Famous Players ordered Eidophor projectors from Philips for nine of their theatres in Toronto, Oshawa, Hamilton, and St. Catherines. This equipment was not available off the shelf in this quantity, and plans were made for the projectors to be delivered and installed in time for the next hockey season in the fall of 1963. On May 28, 1963, a press conference was held in the Telemeter studio to announce this program. On August 6 the first two projectors were delivered at Dorval Airport in Montreal by KLM, and press and photographers were there to kick off a major publicity campaign. The first away game of the Maple Leafs for that season was distributed over the Telemeter cable system and simultaneously exhibited to full houses in the nine theatres on November 7, 1963. Thereafter, these games were exhibited in this manner every week for the rest of that season.

The Telemeter cable network covered an area with 14,000 households to which the service could be provided and had up to 5,600 subscribers during the first year of operation, a saturation level of 40 percent. However, the delays which had unavoidably occurred in adding live productions and sports events had taken their toll, and by 1962 the saturation level had dropped to some 25 percent, although it remained close to that level for the next two or three years. Nevertheless, the costs involved in specially produced live programs and sports could not be supported by the revenue from a few thousand subscribers, and this was the basic reason for exhibiting the hockey games to theatre audiences as well as to the Telemeter subscribers.

It must be remembered that, although the system was operated as a commercial service, it was still intended to be a fairly large-scale field trial. One of its prime purposes was to find the practical market level, and what size of distribution system and saturation level would be necessary for a viable operation. By 1963 sufficient experience had been acquired to indicate that economy of scale was a very important factor, and it was possible to project with reasonable accuracy what size of system would be required at the saturation levels which could be realistically achieved to ensure viability. It was also very clear that if the pay-TV system were sufficiently compatible with CATV so that both services could be provided sharing the same distribution facilities and many of the overhead costs, the operation could be viable at a substantially lower level.

Discussions and studies then began in Toronto and New York on the feasibility of expanding the system in Etobicoke to serve a larger area and adding a CATV service. However, in mid-1963 the Toronto area was not considered to be a viable CATV market, and it was not possible at that time to anticipate the situation which in the following decade turned Toronto into the biggest CATV market in Canada.

In metropolitan Toronto at this time three Canadian TV stations and three U.S. stations located in Buffalo, N.Y., were easily receivable on domestic antennas. There was no Canadian colour broadcasting and therefore few colour TV sets, and high-rise construction in the area, which later interfered with TV reception in many homes, had barely started. Consequently the Telemeter system was a stand-alone pay-TV system with no CATV revenue to share the distribution costs, and no apparent justification for adding these facilities.

Famous Players then started looking for a suitable CATV system in another area to which the Telemeter pay-TV service could be added and had discussions with Dave Campbell who operated Cable TV Ltd., a company serving an extensive area in the west end of Montreal beyond the area served by Rediffusion. Discussions dragged rather fruitlessly through 1964, and by June Telemeter abandoned these plans and arranged to ship their remaining stock of 4,000 subscribers' control units to England, where plans were in hand for a similar pay-TV trial with these units suitably modified for the different electrical standards and coinage.

By the end of 1964 Famous Players was heavily involved in CATV developments across Canada, both in acquisition and expansion of existing systems and the building of new ones, and had accumulated considerable experience of the costs involved, the saturation levels required for viability, and the potential returns on capital invested. This knowledge did not encourage the company to view any plans for expansion of the Telemeter system in the Toronto area with enthusiasm, since it was clear that CATV was a more profitable investment than Telemeter. At the same time Paramount Pictures felt that, after nearly five years, they had achieved the original purpose of the field trial and did not wish to continue picking up any operating losses. As a result the pay-TV service in Etobicoke was discontinued on April 30, 1965.

It was often said subsequently in the media, and generally believed by most people in the television and cable TV industries, that the Telemeter pay-TV experiment in Etobicoke was a failure, and this was often used as an argument against the regulatory approval of any form of pay-TV in Canada. But this is quite untrue, and to use this as an argument against pay-TV was to take it out of context and misuse it.

First, it must be realized that, although run as a commercial operation which did in fact lose money, its essential purpose was as a field trial to gain the very experience which was acquired in its five years of public service. That experience indicated that if it had been expanded to serve a larger market it could have been commercially viable, but it also indicated that viability could be achieved in a smaller market if the distribution costs could be shared with a CATV service in an area where there was a need for this service. However, the very circumstances which led to the choice of the Toronto area for the pay-TV field trial also precluded the addition of CATV in order to achieve that viability on the Etobicoke system.

Second, it must be remembered that this was a one-of-a-kind operation, and equipment had to be designed and manufactured in uneconomically small quantities. For example, the subscribers' control boxes cost almost $100 each landed in Canada from a manufacturer in Chicago and a maximum of 6,000 were produced. If quantities could have been substantially increased and manufactured in Canada, thus eliminating landing costs and spreading the development costs, these units would have been less expensive.

The same cost considerations applied to the production of special programs. The few that were produced demonstrated that there was indeed a market for this type of programming on a pay-by-program basis, but there were not enough of them to be able to get away from the concept that the program schedule was to all intents and purposes comprised of movies. Special productions were costly and more of them could not be justified for only 5,000 or fewer subscribers. Telemeter did manage two seasons of once-a-week hockey, but only because the paying audience could be enlarged substantially by simultaneous exhibition in nine theatres.

There were also some important technical reasons why the Telemeter system was not a commercial success. Basically it was ahead of the technology available. In the decades since this system was installed there has been a veritable revolution in technology and the hardware resulting from it. In 1960 large and expensive vacuum tube amplifiers were still being used in the distribution system and in the subscribers' equipment. Today all that equipment would be solid-state, smaller, cheaper, much more efficient, and less expensive.

Additionally, a significant part of the costs resulted from the use of point-of-sale cash payment. This was selected to permit impulse buying on a per program basis. Today impulse buying can be catered to just as effectively and a lot less expensively by the use of computers and interactive communication. If these had been available then it would have reduced capital costs because the subscriber's control unit would have been a hand-held keypad containing a single chip, instead of a large and costly box containing tubes and a complex coin collection mechanism. Operating costs would have been reduced because door-to-door cash collection would have been avoided. As a commercial venture the Telemeter system in Etobicoke was a failure, but as a large-scale field trial, which was its intended purpose, it was a success.

It was another ten years before any further attempts were made to bring pay-TV to the public, this time in the United States, and this had to await the advent of satellite distribution which brought with it the possibility of achieving the economy of scale so essential to the success of such a service. In Canada, which had provided the practical testing ground in the early 1960s, it was to be even longer, in fact another twenty years, before this more advanced technology was allowed by government regulations to provide a similar public service.

In 1974 Time-Life through a subsidiary, Home Box Office (HBO), commenced a pay-TV service using microwave for distribution to cable systems in the northeastern U.S. within reasonable microwave reach of the New York City area where the programs originated. These systems were mainly in the states of New York and Pennsylvania. Early in 1975 HBO announced an agreement with RCA to use one transponder on its first domestic satellite, Satcom 1, for nationwide distribution of these programs, and this service began that September. This was the first hard evidence of a link-up between two developing technologies, pay-TV and satellite distribution, which between them have revolutionized the communications and entertainment media.

This was a tremendous gamble for Home Box Office because they signed a six-year contract with RCA without having contracts for more than two or three potential users to support it, and without even the assurance that it would work technically, but they had enough experience of the market to know that economy of scale was vital for the success of pay-TV, that this could only be achieved by national distribution, and that this was not practical using the available microwave facilities. By mid-1975 the successful use of satellites for this purpose had been proved with the pay-TV service being used by more than 800,000 subscribers in more than 375 cable systems throughout the United States.

Between 1970 and 1972 several proposals were submitted to the Canadian Radio-Television Commission (CRTC) for a pay-TV service in Canada, either by broadcast or by cable. However, the Commission in its first public announcement on the subject in October 1972 declined to licence any applications because it had concluded that "no submission significantly contributed to the furtherance of the Broadcasting Act objectives," but invited contributions to assist in determining "the best method for a thorough examination of pay-television services as an integral part of the Canadian broadcasting system."

In February 1975 the CRTC issued a position paper on pay-television. The paper described the possibilities and problems foreseen by the Commission, and provided a tentative indication of measures that might be introduced to assure that this service would contribute to Canadian broadcasting objectives. The Commission stated that "unless proposals could be developed which would further the objectives of the Broadcasting Act the impact of pay-television on the Canadian broadcasting system could be highly detrimental." In December 1975, following an extensive public hearing, the Commission issued a policy statement that continued the refusal to issue licences for pay-TV, saying that "the Canadian broadcasting system is still in the process of adjusting to new policies and developments and should be allowed to adapt and absorb these adjustments without the potential disruption caused by the introduction of pay-TV service at this time."

By early 1976 there had been renewed demands from the cable industry for the introduction of pay-TV. There were three main reasons for this renewed interest. First, the early success of pay-TV in the United States had raised expectations of a similar profitable service in Canada. Second, the cable TV industry was faced with decreasing growth rates due to already high penetration levels and, with unused capacity on existing systems, saw pay-TV as a new avenue for the introduction of additional revenue-producing services. Third, cable operators were concerned about the possibility of increasing competition from unlicensed closed-circuit systems in cable franchise areas. A number of these had already been established in hotels in larger centres such as Toronto and Vancouver since 1972.

As a result of this pressure the CRTC held another public hearing in June 1977 which attracted 140 submissions from various interested groups and members of the public. The most strongly represented groups were the industries which could be expected to have a direct involvement in any pay-TV service, including cable, broadcasting, program production, and communications carriers. A considerable proportion of the submissions did not express either support for or opposition to the introduction of pay-TV. Of those that did express an opinion, a surprising majority indicated opposition. The high proportion of briefs which did not express a firm view may indicate that the issues surrounding pay-TV were found to be so complex that a definite assessment of the possible consequences, whether positive or negative, was not attempted.

In March 1978 the CRTC issued its *Report on Pay-Television* based on the 1977 hearings, in which it stated that "no single proposal achieved an acceptable level of commitment to present broadcasting policy objectives and requirements. If the objectives of the Broadcasting Act are to be respected pay-television must be a predominantly Canadian service. It is therefore the Commission's view that it is not possible on the basis of these submissions and the Commission's further analysis to recommend the introduction of pay-television at this time."

Finally in April 1981 the CRTC issued a call for applications for licences for national and regional pay-TV services. Following a mammoth hearing that lasted from September 24 to October 15 and an intensive five-month deliberation period, the Commission approved six of the twenty-seven applications. Two of the applications approved were for general interest services to be distributed nationally, one in English and the other in French, while the other four approvals were for services directed to specific regions, taking advantage of the spot-beam facilities provided on the new Anik C satellites which were not available on the earlier series of satellites.

Pay-TV presents current movies and special events on separate channels from those carrying advertiser-supported programs, and the cable companies then make an extra charge for providing access to those channels. While this has the advantage of offering premium programs which are not available on the commercially-financed services, it has the disadvantage that the choice of material and time of presentation are still left to the program supplier. The viewer has no choice in this selection and is paying for the ability to access the service whether he or she uses individual programs or not, and the only choice is whether to watch the program being offered at the time.

"Pay-per-view" overcomes this disadvantage by allowing the subscriber to decide whether to view a program being offered and then pay only for that program; however, this still leaves the viewer dependent on the program supplier who decides what to offer and at what times. This can be overcome where movie offerings are concerned by providing facilities whereby the viewer can select a movie from a menu of available titles and have it transmitted on request, much as one would select a video from a rental store. This is known as "video-on-demand" (VOD) and is an interactive service as distinct from all other television services, whether on cable or broadcast, in which the viewer is a passive participant choosing between programs being offered but having no choice in the origination of those programs.

Since the program material is selected by, and directed to, an individual subscriber, VOD is in effect the ultimate form of narrowcasting, giving individual subscribers what they want when they want it. The technology for providing such a service is now in place, but its practical implementation requires a substantial increase in the number of channels available; it is this potential new service, on top of the wide range of regular program services now available, that is helping to push the constant demand for more cable channels. Video-on-demand also requires the cable system to provide two-way facilities, either upstream on the cable itself, or via the public telephone network, since the subscriber must now be able to communicate with the program originator to transmit requests for selected programs.

The view has often been expressed that there must surely be some practical limit to the number of channels required on any cable system irrespective of the technical capability of the system to deliver. This is a valid comment in today's programming environment, accentuated perhaps by the fact that on any cable system capable of delivering thirty or more channels there is inevitably considerable duplication in the types of program being offered simultaneously, and further increases serve only to offer more of the same. However, this does not take into account the nature of the new services which will be available when channel capacity is no longer a limiting factor, and video-on-demand is an example of such a service.

The essential feature of VOD is the ability of any subscriber to choose a program from a menu and have it delivered over the cable system at a time of his or her choosing. This can only be accomplished if there are enough channels available to be able to dedicate one to this subscriber for the duration of the chosen program. An example will clarify the effect of this demand in a typical situation.

Assume a distribution node within a cable system's network serves 1,500 homes and two-thirds of these are subscribers. Further assume that, since this will be a discretionary service for which extra charges will be levied, 25 percent of these subscibers will avail themselves of the service. Then VOD will be serving 250 customers in this group. Probability theory suggests that the peak demand for the service might be some 10 percent of these customers, and this would require twenty-five channels to provide this number of individual program selections simultaneously. If half of these 250 customers selected a different program this would require 125 channels, but even this number will be within the ability of the developing technology.

Another possible approach requiring fewer channels which could be applicable to a VOD service devoted to movies could be to offer a number of movies of two-hour duration, each being transmitted simultaneously on a group of channels with the starting times on each staggered at half-hour intervals with repeated showings. This would require four channels for each movie on the menu, but 125 channels could accomodate a menu of thirty-one different movies. A subscriber could choose not only the program but a convenient starting time, and make a selection by channel number without any need for advance ordering and the associated upstream facilities required for this purpose. More conservatively, a typical menu of ten movies could be offered using forty channels.

CHAPTER 6

Growth to maturity: from CATV to cable TV

By the mid-1960s CATV had developed into a budding industry catering to a public need for access to television, the relatively new and rapidly expanding medium of information and entertainment, and a growing desire for a wider choice of TV programs than was available from the stations built to serve their local areas. This desire could only be satisfied by providing the ability to receive distant stations; however, these, by definition, were stations not easily receivable by domestic antenna. This was exactly what CATV, Community Antenna Television or literally an antenna serving a community, was designed to overcome. The desire to increase program choice by adding distant stations to those available locally resulted from circumstances which differed between Canada and the United States.

In the United States by this time all major areas of the country were served by the three national television networks: ABC (American Broadcasting Corp.), CBS (Columbia Broadcasting System), and NBC (National Broadcasting Corp.). These networks, with outlets in every major city, provided a choice between three competing program services which was sufficient to serve the public demand at that time. In these cities there was no incentive to build CATV systems because the distant stations which they could bring in, even if permitted, would be other broadcasting outlets of the same networks and would not increase program choices.

The situation was different in the smaller cities and towns remote from the major areas. Even if these communities were large enough to justify at least one television station, in most cases they were not of sufficient size to justify an outlet from each of the three networks. Consequently, demand for more program choice could only be served by bringing in one or more distant stations not receivable on domestic antennas, and CATV provided this capability. This is why in the first three decades of its development CATV growth in the United States was confined to the many small towns and cities not served by all three national networks and did not develop in the major cities.

By the mid-1970s the public demand for more program choice had developed to the extent that in some of the bigger metropolitan areas there was a large enough market to justify the cost of building further television stations independent of, and competing with, the established networks. In the largest markets these extended the viewing choice to as many as six or more stations. But there were many cities which, while able to support three network outlets, were not large enough to provide the additional advertising volume required to support the cost of one or more independent stations.

Here again CATV could have provided the answer but was inhibited by FCC (Federal Communications Commission) regulations which, as a means of limiting destructive competition with its licencee broadcasters, would not permit the reception and distribution by cable of stations outside their authorized coverage area. This effectively limited the growth of cable services to areas which were inadequately served by local off-air transmitters and could reasonably justify investment in cable distribution to make all the networks available. In fact, in 1975, although there were some 3,400 systems in the United States serving approximately 10 million subscribers, the average size of these systems was less than 3,000 subscribers, and 98.5 percent had fewer than 2,000.

As a result the development of CATV in the United States, concentrated as it was in the small town markets, barred by FCC regulation from the major cities, and with inadequate access to capital financing, stagnated through the late 1960s and early '70s. It was effectively limited to the smaller communities in which reception of distant stations was permitted because of inadequate local coverage by the networks, and no cable systems were built in the larger cities and metropolitan areas.

The "satellite era" began in 1975 with the first nationwide transmissions of pay programs by Home Box Office (HBO). The potential for this new form of television distribution had been demonstrated for the first time at the annual convention of the American National Cable TV Association in Anaheim, California, in June 1973 with a satellite link-up between Anaheim and Washington DC. At this time there were no U.S. satellites which could carry television programming, and arrangements were made through the FCC in the United States and the Department of Transport in Canada to use Canada's Anik satellite for this special demonstration. Following HBO's first transmission in September 1975 it was not long before many other program services were also using satellites for national distribution. This effectively bypassed the FCC's prohibition on distant station reception, since, for those program services distributed by satellite, there was now no such thing as a "distant station." This gave rise by 1980 to an intense scramble for franchises to build cable systems in just about all the major cities in the U.S. which had been affected by this prohibition.

In the United States permission to build a cable system is given not by a licence of the federal government, as in Canada, but by a franchise from the municipality, granted under the authority of a bylaw giving the applicant permission to use the public rights-of-way. The potential increase in television program services made possible by the use of satellite distribution, together with the public demand for these services, was sufficient to justify the investment of the millions of dollars required to build any cable system capable of serving a population of several million. In fact the possibilities were so attractive that applications for franchises in all the major markets flooded in; in some cases there were as many as ten or a dozen received for each. By 1983 franchises had been awarded in just about every major metropolitan area in the United States, and very sophisticated cable systems were under construction in Detroit, Denver, Milwaukee, Tuscon, and many other cities.

In Canada the competitive development of television broadcasting has been much more closely controlled by the federal government and the natural forces of free enterprise have not been allowed free reign. When television broadcasting commenced in 1952 responsibility for its initial development was given to the Canadian Broadcasting Corporation (CBC), a government corporation responsible to Parliament. Although the CBC obtains revenue from commercial advertising this only meets part of its operating costs, and the balance of these costs and its entire capital expenditure is subsidized from the federal treasury. In order to develop a television service across Canada as rapidly as possible the CBC was authorized to build stations in the major cities and to create a program network linking those stations. Private broadcasters were licensed to build stations in other areas but were required to affiliate with the CBC network and carry a substantial amount of that network's progamming. No competitive stations were allowed during this build-up period from 1952 to 1960, so that by the end of this time some 95 percent of the total population had access to a single Canadian television service.

In 1960 the government, through the Board of Broadcast Governors (BBG) which had been formed by the Broadcasting Act of 1958 as the new regulatory agency, decided that, since this first aim had been achieved, a second competing service could be introduced to provide a program choice. Second stations were then licensed in the larger centres to private broadcasters, and these were linked together by a second network, also privately owned and operated.

This plan for the orderly development of broadcasting in Canada was very rational, giving priority to the availability of a television service to as many Canadians as possible before making a choice of programs available as a second priority. Unfortunately it did not take into account two facts of life in Canada. First, it is basic human nature that, once the novelty of television has worn off, the viewer who has only one program available wants a second, and the viewer with two wants a third, etc. Second, the population of Canada is strung along the international border and more than 50 percent of Canadian homes are more or less within reception distance of television stations in the U.S. border states.

It was inevitable that the government-imposed limit of one station per market and one national network should have created a strong urge to obtain a wider choice of programs by looking to U.S. stations, and this urge was just as strong in the big cities as it was in the smaller towns. Hence the fact that all the roof-top antennas in cities close to the U.S. border — like Toronto, Oshawa, Hamilton, Winnipeg, and Vancouver — were pointing, not at the local Canadian stations, but across the border at stations broadcasting the U.S. networks. Where reception of these stations was less than satisfactory on domestic antennas then CATV systems were built to improve this quality.

Another factor contributed to this demand for wider program choice. Part of the BBG's mandate was to encourage a distinctive Canadian culture in television broadcasting. The Board chose to do this by making it a condition of every television license that at least 55 percent of the program schedule had to be "Canadian in content and character." This legislation, limiting foreign program content, effectively barred many popular U.S. network programs which might have been aired by Canadian stations and fed a growing interest in obtaining these directly from the stations over the border which were carrying them.

For these reasons CATV had developed in Canada even more rapidly than in the United States. By 1966 there were some three hundred systems in Canada serving 310,000 subscribers, a level of penetration more than twice that in the United States.

The necessity for International Telemeter to build a cable system in Canada to provide a suitable vehicle for a large-scale field trial of their pay-TV system, and the tie-in with Famous Players Canadian Corp. for this purpose, brought the movie theatre company into the CATV field. The initial intention to build a cable system for pay-TV distribution in

London, Ontario, brought Famous Players into contact with Ed Jarmain, and later, after this trial had been moved to the Toronto area, into partnership with him in expanding his existing CATV system in London.

This early experience indicated to Famous Players that, quite apart from the success or otherwise of the pay-TV experiment, CATV itself could be a major public service of the future and a potentially profitable form of investment. As a result Famous Players started looking at other communities in Canada where domestic television reception was limited, but their locations were such that it might be possible to receive and distribute by cable one or more American TV stations broadcasting across the border.

The first of these was Cornwall, Ontario, where a small CATV system, owned and operated by the Bertrand family, had access to two television stations in the United States — one in Plattsburgh, New York, and the other in Burlington, Vermont — which were outlets of two of the U.S. networks. Bertrand Senior was anxious to retire, and his son, who was managing the system, was interested in acquiring a partner who could provide the substantial financing required to expand.

At that relatively early period in the development of CATV — less than ten years from the time when the first CATV systems were built in Canada — obtaining financing was one of the biggest problems facing the fledgling industry. Cable TV is, and always has been of necessity, a capital intensive industry, since a headend must be built and several miles of cable and line equipment installed before the first subscriber can be connected and revenue starts flowing. Banks and other lending institutions were not prepared to finance private investors in developing CATV systems because the future of this new service was too uncertain, the past was too short to produce any reliable payback history, and in their opinion there just was not adequate security.

There were therefore a number of systems across Canada that had been started by small entrepreneurs with largely private funds which were unable to expand to meet a growing demand because they could not arrange sufficient borrowing and could not finance adequately out of cash flow. Famous Players was a company with financial resources available for investment and an existing interest in cable through its involvement in the Telemeter pay-TV experiment. This involvement led the company to the realization that CATV could satisfy a growing public demand for more television and could become a major revenue source for the future.

Arrangements were completed in February 1961 for the purchase of a half interest in Cornwall Cablevision. This was the first acquisition of many across Canada which, by the end of the decade, made Famous Players the largest multiple cable system owner in North America. In fact, by the time the Cornwall purchase was completed discussions were also proceeding with Fred Metcalf, who was operating a successful system in Guelph, Ontario, to go into partnership to build a system in the nearby Kitchener-Waterloo area. In August 1961 a company was incorporated for this purpose, Grand River Cable TV Ltd., which included Carl Pollack as a shareholder and director. Pollack was the owner of the local television station CKCO-TV and also owner of Electrohome Ltd., a manufacturer of consumer electronic products based in Kitchener, and at that time one of the few Canadian companies producing television receivers.

Famous Players was also pursuing similar investment interests in the west, where over the next two or three years they purchased equity participation in existing systems in Port Alberni on Vancouver Island, Powell River, B.C., and Estevan,

Saskatchewan. New companies were also set up with the co-operation of local shareholders to build systems in Weyburn, Saskatchewan, Lethbridge and Medicine Hat, Alberta, Winnipeg, Manitoba, Barrie and Orillia, Ontario, and the twin cities of Port Arthur and Fort William, Ontario (since amalgamated into Thunder Bay). Also in 1963 Famous Players took over Hamilton Coaxial, a company operating a system serving part of Hamilton, Ontario. This company had excellent potential but was in serious financial difficulties.

While these ventures were progressing, back in Toronto, Famous Players' home town, another seed was being sown which was to develop by the 1980s into what was probably, at that time at least, the biggest concentration of cable television subscribers in the world. In May 1961 Peel Village Developments Ltd., a major residential developer, approached Famous Players for advice on a problem they were facing. The company was planning the development of a large subdivision of single-family homes on the outskirts of Brampton, a rapidly growing suburban area about twenty miles from the centre of Toronto. This subdivision, to be known as Peel Village, was to be a model community, and, unlike most of the contemporary developments which were marred by unsightly utility poles and overhead wiring, its esthetic attractions were to be enhanced by placing all the services, including telephone and power, underground.

Peel's problem was that, after going to this trouble and expense, they knew full well that once the homes were occupied almost the first thing the new owners would do would be to erect television antennas on their roofs to receive the U.S. TV stations in Buffalo, New York. This they were determined to avoid because it would spoil the visual effect they were planning for.

Famous Players told Peel that the answer to their problem was a cable system connecting all the homes to a central antenna, that is a CATV system, and that this cable could be installed underground with the telephone facilities. Famous Players, through their subsidiary Trans Canada Telemeter, already had a partial lease agreement with Bell Telephone for the cables installed in the Telemeter service area in Etobicoke only a few miles away, which could be extended to cover the Peel Village area of Brampton.

Famous Players agreed with the development company to install cable TV facilities into every home during construction. Bell installed the main distribution cables with their telephone plant in rear lot easements, and Famous Players installed the electronic equipment used with this cable in pedestals placed by the telephone company. Famous Players also ran a buried drop cable into each home from the pedestals during house construction, extending this by internal wiring to locations as required by the owner after purchase and occupation. In turn Peel Village Developments paid the company an installation charge for each drop installed and wrote a prohibition on external antennas into the deeds of each property to achieve the desired esthetic environment. This gave a paid-up installation fee before the owner moved in and almost guaranteed 100 percent saturation right from the start. Almost but not quite, because a few owners refused to pay for cable service and each installed an antenna internally in the roof space. However, not many people were prepared to accept this alternative because the stations in Buffalo were too distant to be reliably received on an antenna capable of being installed in such a limited space.

It was not long before the success of this system in avoiding the eyesores of utility poles, overhead wiring, and the inevitable forest of roof-mounted TV antennas attracted the attention of other developers. Famous Players was soon approached by others planning new residential construction in the metropolitan Toronto area to incorporate cable TV facilities in their homes as part of the development. To cater to this new market, which was an offshoot of the Telemeter experimental system in Etobicoke but quite separate from it, Famous Players incorporated a new company, Metropolitan Cablevision, as a wholly-owned subsidiary of Trans Canada Telemeter. By 1963 new subdivisions were being cabled by the company in several of the developing bedroom suburbs around Toronto — Etobicoke, North York, Scarborough, and Mississauga, a rapidly growing suburban area to the west of Etobicoke then known as Toronto Township.

Buried construction was relatively new to the CATV industry. Indeed it was also new to the utilities too, and all involved had to learn many lessons from practical experience. One of the biggest headaches was one which many cable TV operators have since experienced and probably still do — the placing of cable in ground prior to final grading. This applied to the distribution cables as well as to the drops, since unlike most subdivisions today, the cables were in rear lot easements and not in the road allowance where grades are established early in the construction program.

Frequently, one would see pedestals a couple of feet above ground level literally supported by the cables terminating in them, or alternatively sitting in a hole in the ground with the top level with the surface and just waiting for the first heavy rain to become flooded. The other problem which became prevalent in the early days with this type of construction was the damage caused to cables by owners digging, especially when fences were being erected and trees and shrubs planted. Here again there was not the awareness among homeowners that there is today of the prevalence of buried services.

In the early 1960s Toronto itself was not considered to be a viable CATV market. The subdivision developments had been successful in spite of this because they provided an esthetically acceptable alternative to roof-top antennas under circumstances in which the latter could be controlled. However, this situation was peculiar to new housing developments and could not be expected to succeed in existing built-up areas where domestic antennas were already an established feature of the environment.

By early 1965 Famous Players was making plans to close down the Telemeter pay-TV operation. It was clear by then that it could continue as a viable operation only if the size of the market could be substantially increased or the cable network could also be used for CATV distribution and so share the overhead costs. However, the latter was not practical in the face of established roof-top antennas which would severely limit any demand for the service. Famous Players was by then convinced that capital expended on CATV systems in other areas where there was a need would be a much more promising investment than any extension of the Telemeter service in the Toronto area. Paramount Pictures, for its part, had learned all it needed from the five years the system had been operating in Etobicoke and were not prepared to continue financial support.

There was another development brewing which was to change this situation dramatically. Although colour broadcasting had been available for some years in the United States, it still was not permitted in Canada by the Board of Broadcast Governors (BBG), the broadcasting regulator at that time and the predecessor of the Canadian Radio-Television Commission (CRTC) which replaced it in 1968. In late 1965, it was known within the broadcasting industry, though not well-known publicly, that the BBG was considering allowing Canadian broadcasters to convert to colour by the end of 1967. Off-air colour reception is much more critical of imperfections such as weak signals, reflections, etc., which make satisfactory colour reception much more difficult in urban areas than the reception of black and white. These problems can be overcome by the use of CATV with a central antenna whose location can be chosen for freedom from these imperfections. This alone, however, would not have been adequate to ensure that Toronto would be a sufficiently viable CATV market to risk millions of dollars in investment if it had not been for another change which was also developing at that time.

Up until the early 1960s Toronto had always been typically a city of single family dwellings, unlike Montreal in Quebec, where a much larger percentage of the population occupied apartments and similar multiple dwellings. Around 1964 this situation began slowly, but later with increasing momentum, to change. First, there was a trend to high-rise office construction in the downtown area, one of the earliest buildings of this type being the Toronto-Dominion Centre rising to an almost unheard of fifty-four storeys! This was followed, or possibly accompanied, by a similar trend in residential construction towards high-rise apartment towers rising to as much as twenty storeys or more. Furthermore, this high-rise construction was not confined to the downtown area but started sprouting all over the suburbs.

It was the coincidence of these two developments — high-rise construction and the switch to colour broadcasting — that triggered the possibility of CATV being a viable service in the established residential areas of Toronto. High-rise buildings would not only block reception of the popular distant U.S. TV stations seventy-five miles away in Buffalo, to which practically every domestic antenna was pointing, but would also cause multiple reflections on the local Canadian TV transmissions in some areas, and the effects of these would be greatly accentuated as viewers gradually turned to colour reception.

With these considerations in mind and a potential deadline of late 1967 for colour transmissions from the Canadian stations, Famous Players decided to test the market to determine whether they would be justified in proceeding with cable development in advance of the deadline in order to be in a position to offer CATV service as a solution when owners started to convert to colour receivers. Fortunately, there was a potential test bed immediately available. When the Telemeter operation in Etobicoke closed down in April 1965 the company had an outstanding contract with Bell for 103 miles of cable in place at a monthly rental of $1,000 with four-and-a-half years still to run. Although the contract could be cancelled without penalty it represented an investment of some $200,000, only a little more than half of which had been written off. This network covered 14,000 homes, with drop cables still in place in more than half the homes, representing a replacement value of $150,000.

It was decided that Famous Players would make use of part of this wired area for a CATV market test, although it was realized that it would not necessarily be the most promising market in the Toronto area. Apart from the ability to use the existing cables, an important factor influencing this choice was the fact that Metropolitan Cablevision had a licence authorizing the company to provide CATV service in Etobicoke. The receiving station included in this licence had been established to serve two of the new residential subdivisions and was conveniently located about two-and-a-half miles by cable from the nearest point on the Telemeter network.

The area chosen for the experiment covered 1,800 single family homes with approximately thirteen miles of distribution cable. Famous Players arranged with Bell to continue the lease on this portion of the cable, but to suspend all monthly payments on the unused ninety miles with an option to resume the lease on any or all of this at any time within twelve months from February 1966. In the test area all Telemeter equipment was removed and all subscribers' drops disconnected. The whole of this cable was then tested to ensure its suitability for the much higher frequencies involved in CATV. Some additional cables were added to meet the planning requirements of a CATV system, including two-and-a-half miles of trunk cable needed to reach the receiving site, and new CATV amplifiers and associated equipment were installed. With these changes and additions required to the existing cable system it was not expected that it could be made ready for the new service before the end of 1965.

In the meantime, Famous Players initiated extensive market surveys to determine whether there might be another area in metropolitan Toronto more suitable for this service. These surveys revealed that there was a small percentage, some 10 to 15 percent of the respondents, who would be prepared to pay for a cable service because they were dissatisfied with antenna reception for various reasons including cost, hazard, damage to the roof, electrical interference, need for replacement, and, by no means least, a desire to receive the new Buffalo non-commercial PBS station not easily received by the average domestic antenna.

Other surveys had indicated that in 1965 less than 1 percent of the 500,000 TV homes in metropolitan Toronto had colour TV sets, that these were concentrated almost exclusively in the higher income areas, and that this figure would increase to at least 20 percent by the end of 1968 if Canadian broadcasters were allowed to introduce colour before the end of 1967. When this estimate was taken into account along with certain technical considerations it led to a very optimistic prognosis for CATV. Colour demands much higher technical standards for acceptable reception, and existing antennas producing a degraded but tolerable picture in black and white would in general be unsuitable for colour, particularly where the viewer's expectations would be much more critical after he or she had paid between $600 and $1,000 for a new TV set.

Consequently, it appeared very probable that the advent of Canadian colour broadcasting, coinciding with the new and intensive developments in high-rise construction, would create a CATV market in Toronto whose growth would be closely linked to the rate of replacement of old receivers by colour receivers, providing the service was available before these sets were purchased. Since the purchase of new receivers, which were appreciably more expensive than the black and white sets they would replace, would most likely commence in the higher income areas these became the preferred targets for early CATV development.

Later in 1965 another factor was injected which could have drastically changed the estimates of CATV growth — an announcement that CKVR-TV, broadcasting on channel 3 in Barrie sixty miles north of Toronto, was planning to move its transmitter to Palgrave less than twenty miles north of the metropolitan area, making it virtually a third Toronto station. This would have caused serious interference with reception of the Buffalo network stations on the adjacent channels 2 and 4. Although the BBG eventually rejected the move after an appeal to the federal cabinet, it certainly appeared in 1965 to be another reason for seriously considering the development of CATV in Toronto.

In June 1965 Famous Players sent Bell a letter of intent requesting the provision of CATV cable facilities throughout the whole of metropolitan Toronto plus Toronto Township, an expanding suburban area to the west, now incorporated as the City of Mississauga. This was a normal procedure used by Bell at that time to establish a construction priority in the areas concerned. So long as there were no other applicants, this intent could be implemented more or less at leisure. However, if another company informed Bell that it was prepared to place a firm order the telephone company would give the first applicant thirty days to confirm its intent with a definite commitment. If the option was taken up, then Bell would inform the second applicant that it was prepared to build the system after it had completed construction for the first. There was in fact no exclusivity exercised by Bell, but experience had shown that it was most likely that the second applicant would withdraw because of the substantial competitive advantage the first comer would have, especially in an untried and uncertain market.

With all these considerations in mind, in February 1966 Famous Players made the following decisions:

1. to exercise the option on the rest of the Telemeter network in Etobicoke during 1966 with the object of having it in service as a CATV system by the end of the year serving approximately 10,000 single family dwellings;
2. to proceed immediately with plans to build a CATV system in the area bounded by Highway 401, Don River, Bayview, Eglinton, and Avenue Road, serving 13,000 homes, for completion by the end of 1966;

3. to survey the metropolitan area as a whole to determine the areas most likely to benefit from CATV in order that they may be given priority in future plans;
4. to be prepared, if necessary, to proceed with construction in the order of priority indicated by the survey, at a rate dictated by Bell's construction capacity.

These decisions represented a substantial increase in the scale of Famous Players' involvement in cable operation in the Toronto area, and in order to more effectively handle this a separate company was incorporated under the name of Metro Cable TV Ltd. The new company applied for and received licence approvals for all the receiving stations in the metropolitan Toronto areas previously licenced to Famous Players and had them amended where necessary to include authorization to serve the whole of the metropolitan area, plus certain contiguous areas.

Planning then proceeded for the areas defined in the decision, starting with Hogg's Hollow, located immediately east of Yonge Street below Highway 401. This is an area where there were some three hundred expensive homes situated in a deep ravine formed by the Don River in which television reception, even of the local stations, was very difficult, and there was little doubt about the need or demand for CATV service. From here the pre-construction planning proceeded progressively south as far as Eglinton Avenue.

The market survey covered some 46,000 homes in north and central Toronto as far south as Bloor Street, and included a further 9,000 homes in Etobicoke. From this survey, together with field investigations, an area comprising 35,000 single family homes was selected for further development as a potentially viable CATV market. This included such relatively affluent communities as Forest Hill Village, Rosedale, and the area around Upper Canada College, where early penetration of colour TV sets was most likely and which were already experiencing problems with some distant station reception. In fact 30 to 50 percent of the respondents to the survey in these areas had shown interest in a cable service.

In August 1966 Famous Players extended the system to these areas, and in February 1967 a further expansion covering another 20,000 single family homes was approved and orders placed with Bell for installation of the cable. In the early projections of market potential, apartments had not been included because practically every multiple dwelling building of any size already had a master antenna system for use by the tenants. However, as construction of the cable system proceeded it appeared the projections had been unduly conservative, since the company had already been approached by a number of building owners who were in need of improvement to their television reception. Inclusion of apartment dwellers as subscribers then increased the potential in the planned areas by well over 20,000 homes.

Famous Players expected that if it was proved there was a viable market for CATV in Toronto, there would be considerable competition from others interested in these developments and that timing would be all-important. In fact by February 1967 the Department of Transport had issued nine CATV licences in metropolitan Toronto in addition to those issued to Metro Cable TV.

Most of these were for restricted areas or areas of no immediate concern to Metro Cable. But two of them were for substantially the whole of the metropolitan area, and a third covered the part of Etobicoke where Metro still had an option with Bell on the original Telemeter network, together with contiguous areas in the Borough of York east of the Humber River. The latter licencee, Hosick Television Co. Ltd., later acquired by David Graham and renamed Graham

Cable TV Ltd., had already started operations in a small section of York centred around Jane Street. Famous Players did not believe that this company constituted any serious competitive threat in the immediate future, but Hosick was in a good position to move quickly into the former Telemeter area if Metro failed to take up its option, and in those circumstances Bell would probably have made the existing cables available to Hosick.

The other two licences covering the metropolitan area were issued to David Graham on July 13, 1966, and York Cablevision Ltd. on January 5, 1967. York Cablevision appeared to be a more serious and immediate threat to Famous Players' plans for expansion, since it was reported that York had applied to Bell for cable construction in all areas not already ordered by Metro. This was probably true because Bell ran scared. Although having a complete monopoly of the telephone service, Bell said it could not possibly accept orders which would allow two companies to have a virtual cable monopoly between them throughout the whole of metropolitan Toronto. At that time there was no CRTC or equivalent regulatory agency determining cable TV licence areas, and cable licences issued by the Department of Transport (DOT) were not exclusive. Bell decided it would resolve this dilemma by accepting orders from any one company for a maximum of one million feet of cable and would accept further orders only after half that quantity had been installed.

By this time the amount of cable ordered by Metro from Bell "coincidentally" totalled one million feet. Since it was going to be some time before Bell could complete the installation of half of this, Metro was prevented from expanding the initial area any further at that time. In the meantime York had decided on their preferred area requiring this amount of cable and placed their order. Before the end of 1967 everybody else who was interested had placed their orders for up to one million feet, designating the areas they were interested in on a first-come-first-served basis, so that the latecomers had to be satisfied with the less choice areas.

Probably only Bell knows the exact order in which the choices were made, but Ted Rogers, Maclean Hunter, Barry Ross, Geoff Conway, and others had all applied within a few months. By the end of 1967 there were few areas of metropolitan Toronto which had not been spoken for. The cable TV boundary map of Toronto, which looked like a gigantic jigsaw puzzle and which the CRTC and many others subsequently deplored as cutting across municipal boundaries and having no consistency, was set for at least the next decade. In later years many people were puzzled at the apparent aimlessness of the licence area boundaries and the fact that some companies were operating in several areas which were not contiguous. Many blamed the CRTC, but it was no fault of theirs; the Commission was not even in existence as the regulatory agency until 1968 when its first mandate was to issue licences under the new Broadcasting Act to all existing holders of DOT licences.

The jigsaw puzzle started with the boundaries originally drawn for Metro Cable TV, based partly on the results of the market survey in early 1966. The imposition by Bell of the one million feet quota then allowed the division of territory to be expanded in a descending order of preference until the whole metropolitan area had been allocated, and licences were issued by the Department of Transport to conform with these boundaries. It was not until the late 1970s that the increasing importance of community programming and the desirability to identify communities with municipal boundaries led finally, at the initiative of some of the operating companies themselves rather than the regulatory agency, to some exchange of areas and a move towards rationalizing some of the licence boundaries.

This process by which a complete metropolitan area was developed for cable television service has been described in some detail because it was the first of its kind in North America and was the precursor of the many urban cable systems in operation in all major cities today. By 1968 similar systems had been developed in most of the larger cities in Canada, but none were as extensive, covering most of a metropolitan area, as the group then under construction in metropolitan Toronto.

Certainly nothing as extensive had been developed to serve major cities in the United States, because of the ready availability of the three national networks and the FCC prohibition on bringing in distant signals. It was not until after the introduction of television program distribution by satellite in 1975, which virtually made the notion of "distant station" meaningless, that the rush to build cable systems in many of the largest American cities began and gave the cable TV industry the impetus to achieve the position it now occupies in the broadcasting and communications infrastructure.

Clearly the progression from CATV, providing a needed service to small communities inadequately served by local broadcast transmitters, to Cable TV, providing more sophisticated services to both small and large communities, was prompted by different factors in the United States and in Canada. In the U.S. it was the introduction of nation-wide satellite distribution in 1975 which overcame the regulatory limitations on "distant station" reception and triggered unbounded growth. In Canada it was primarily the introduction of colour television in 1967, and then later the need for a sophisticated antenna system to receive the many additional services being offered by satellite which had the same effect. In both countries cable TV had become by the early 1980s an increasingly important part of the communications infrastructure and was in many homes almost as ubiquitous as the telephone.

CHAPTER 7

Government regulation, trade associations, and the CCTA

There are very few industries that do not find it necessary to establish some form of cooperation among the various companies comprising the industry for good and legitimate reasons — combatting discriminatory regulation, taxation, unfair competition, upgrading industry standards, training and education, etc. This is particularly the case with the cable television industry since, as a form of public communication, it is closely linked with the older and more established industries of broadcasting and telephone communications, and therefore liable to attract the early attention of legislators and regulators. Indeed it is generally some threat in these areas to a young and struggling industry that initiates the drive to organize into a more cohesive whole, and the regulatory climate in which such an industry exists has a considerable influence on the form and rate of its development.

In the United States it was a threat to impose an 8 percent federal excise tax on all revenues that first brought the fledgling CATV industry together. The year 1949 saw the first CATV system in the United States operating as a business venture with regular service charges to its customers, and by 1951 the early systems in Pennsylvania were attracting considerable publicity. Unfortunately, they soon attracted the attention of the Internal Revenue Service (IRS), which quickly recognized a new taxing opportunity, and in that year imposed an excise tax on two of the systems operating in Pennsylvania.

In those early days CATV, with no track record and using an untried technology, had no access to debt financing. Furthermore, it was then, and still is, a capital intensive industry, since substantial expenditures in the installation of cable and equipment are required before even the first subscriber can be connected and regular income generated. The early systems found it necessary to make a substantial installation charge as each subscriber was connected in order to produce the cash flow needed to finance continuing construction. In imposing their excise tax on these systems the IRS considered these charges to be part of income and taxable as such rather than a contribution to capital cost.

This problem brought together a small group of operators at the Necho Allen Hotel in Pottsville, Pennsylvania, on September 18, 1951. There was a long discussion on what cash receipts were taxable, and whether standard depreciation rates and standard accounting practices could be adopted. It was decided there was a need for an organization that could hire the best tax consultants and lawyers on a cooperative basis to tackle these immediate problems. This is precisely how trade associations start — cooperating to face a common problem, and this one was true to form.

The first organizational meeting of the group occurred in Pottsville on September 26, 1951, with eleven of the nineteen charter members in attendance. By January 16, 1952, when the first formal meeting was held, the group had been incorporated as the National Community Television Association Inc., and Marty Malarkey of Pottsville, PA, was elected as the first president with a supporting board of seven directors. By this time CATV had spread beyond Pennsylvania into other states and thus justified the new association in becoming national in scope. Although the majority of its members were systems in Pennsylvania, one of the seven directors was from Memphis, Tennessee, a second from Laconia, New Hampshire, and a third from Carmel, California.

The first annual meeting or convention of the NCTA took place in Pottsville, PA, on June 9, 1952. By this time the association had recruited thirty-five member companies out of a total of approximately ninety-five systems operating in the United States. Of these, twenty-four were in attendance at the convention, together with twelve representatives of manufacturers and news services, and there was a total attendance of seventy-two including guests. At this first convention Marty Malarkey was re-elected president, an office he continued to hold for the next four years until he was succeeded in 1956 by Bill Daniels of Texas.

The Park Sheraton Hotel in New York hosted the second convention on June 8, 1953, the first general meeting held outside Pennsylvania. By then the asssociation had retained a Washington law firm to represent them on the still unresolved tax question and on other legal matters. E. Stratford Smith representing this firm attended his first board meeting as NCTA's legal counsel on December 1, 1953. Smith was by no means new to the problems and promises of the cable television industry since in 1948, as one of the staff of the Federal Communication Commission, he was involved in discussions as to whether the pioneer systems then being built should be regulated by the federal agency. Indeed it was Strat Smith who was credited with coining the term "CATV" because he got tired of repeatedly writing Community Antenna Television.

The third convention of the NCTA was again held at the Park Sheraton Hotel in New York on June 14 and 15, 1954. At this convention members decided that the association should be based in Washington D.C., and the next meeting of the board of directors was held at the new offices at 710 Fourteenth Street N.W. By this time Strat Smith was closely involved not only in the legal problems of the NCTA but also in its administration, and at this meeting he took on the additional responsibilities of executive secretary.

By early 1957 the industry, and with it the association, was growing to the point where it was impractical for Strat Smith to continue to function as both general counsel and executive secretary and it became necessary to hire a full-time administrator. By the time of the next convention on June 4 and 5, 1957, in Pittsburgh, Ed Whitney had been hired as the first full-time executive director of the NCTA.

At this time similar developments in CATV were taking place in Canada. Because the technology and the business problems involved in building and operating these systems were so new, Canadians had a clear common interest with those undertaking similar developments in the United States. From the time of the first NCTA convention in New York in 1953 several Canadian companies had joined the association, and an increasing number of Canadians were regulars at the conventions.

By 1956 the advantages of their own national association were becoming apparent to these Canadian members, and so were the problems in Canada with telephone company relations and potential government regulations which required a cooperative approach but which clearly could not be handled by any non-Canadian organization. On September 27, 1956, a meeting was arranged in Montreal to discuss such a Canadian organization, attended primarily by operators in the province of Quebec, which had the largest number of systems. This was followed on October 22 by a similar meeting in Toronto for operators in Ontario. It was decided that a Canadian association should be formed and an organizing committee was elected which met several times between January and May of 1957. Then on May 29 in Montreal a board of seven directors was elected, with Fred Metcalf of Guelph, Ontario, as president, and Ken Easton of Montreal, as secretary, to apply for Letters Patent to incorporate the association and organize it until a first formal meeting of the members.

The incorporation took considerably longer than had been anticipated due to difficulties arising from the choice of a name acceptable to the association, the Department of the Secretary of State, as the federal registration agency, and the Department of Transport, which was the licencing agency for CATV at that time. The association was anxious to use the name "National Community Television Association of Canada," thus identifying themselves with the NCTA in the U.S. yet providing a clear national distinction. The Department of Transport objected to this because they considered the term "community television" applied, or could be construed as applying, to low-power broadcast transmitters or translators serving specific communities. A compromise was finally arrived at by including the word "antenna," thus adopting the somewhat cumbersome name "National Community Antenna Television Association of Canada" or NCATA of Canada. Letters Patent incorporating the association under federal charter were granted on August 23, 1957.

Eleven years later, in 1968, the association decided to broaden the reference in the name to get away from the somewhat narrow concept of a community antenna. By that time CATV systems were beginning, under pressure from the regulator, to include channels carrying non-broadcast material, more specifically community programming originating within the system. Revised Letters Patent were therefore applied for and issued, changing the name to "Canadian Cable Television Association" or CCTA, and formalizing references to the industry from "CATV" to the more descriptive "Cable Television."

The purposes and objectives of the association, as set out in the incorporating Letters Patent, were:

(a) to promote the interests and conserve the rights of those engaged in the reception, amplification, and distribution of video and audio television signals and in the business of selling or renting and installing special receiving antenna, television amplification, and distribution devices and television receiving sets and associated equipment;

(b) to protect its members against unbusinesslike methods in such businesses or allied businesses;

(c) to foster such businesses and reform abuses therein where they exist;

(d) to diffuse accurate information among its members; and

(e) to secure uniformity in usage, custom, and trade conditions.

The first convention of the association was held on October 18 and 19, 1957, at the Alpine Inn, Ste. Marguerite, Quebec, with fifty-nine active members and seven trade members present. The program of action presented to the members at this meeting indicates the broad range of problems which were facing the pioneer CATV operators, all of which have been subsequently resolved, in general, to the benefit of the industry. They included: Department of Transport licencing regulations and requirements; CBC and Department of Transport policy towards the redistribution of U.S. broadcast programs; licencing of microwave links for this and similar purposes; Bell Telephone Co. rates and services; joint use of poles and relations with public utilities; and industry attitudes towards the movie industry and pay-TV.

At that time, in 1957, the Board of Governors of the Canadian Broadcasting Corporation was the licencing authority for all broadcasting. This was an unsatisfactory situation to practically everyone concerned, since the CBC also had the responsibility of providing a national broadcasting service. By having the authority to issue licences to private broadcasters the CBC was in the position of being judge and jury to their own potential competitors. Thus, they licenced all private broadcasters and then competed with their own licencees.

The government finally recognized this anomalous situation and corrected it by the 1958 Broadcasting Act. An independent Board of Broadcast Governors (BBG) was appointed as a regulatory commission, divorced from any broadcasting operations, with regulatory authority equally over the CBC as well as private broadcasters. The relations of the BBG and the CBC were those of a regulatory body to an operating licencee, and the regulations of the Board which applied to the private licencees applied equally to the CBC.

However, CATV had never been included as part of the broadcasting system since it was considered simply a passive distributor of off-air signals operating at the receiving end of the chain rather than at the broadcasting end, replacing the viewer's own antenna. At this time this was very largely true. The raison d'être for these early systems was simply to improve reception of stations which were available on domestic antennas but not too easily received due to distance or intervening obstructions, and they were limited by their DOT licences to receiving these stations at locations within ten miles of the community they were intended to serve.

Any government control over these systems considered to be necessary was therefore largely of a technical nature and was administered by the federal Department of Transport by means of a very simple licencing procedure. Indeed the procedure was so simple that an application accompanied by a payment of $25 was sufficient to bring a licence almost by mail order, and any conditions attached to that licence were extremely minimal, at least by today's standards.

The licence conditions were spelled out in the General Radio Regulations, Part II, made under the Radio Act, and consisted literally of the following:

> "Commercial Broadcasting Receiving Station licences may be granted by the Minister for stations to be operated for gain for the reception of broadcasting. These licences are subject to the following conditions:
>
> "(a) the licenced stations shall receive signals from the broadcasting stations only that are specified in the licence;
>
> "(b) during the scheduled hours of operation of a Commercial Broadcasting Receiving Station broadcasting received by that station shall not be altered or curtailed in any way except by agreement with the broadcasting station;
>
> "(c) the radio receiving apparatus, including the distribution system, amplifiers, and other devices in use, shall conform to the technical conditions prescribed in the licence."

The technical conditions prescribed in the licence were spelled out in Radio Standards Specification No. 102, first issued by the Department of Transport Telecommunications Branch on March 5, 1956. The intent as defined in the preamble, was "to establish minimum standards for the performance of Community Antenna Television Systems." However, as a technical standard it was very elementary and largely subjective in form.

RSS102 specified limits on the field strength of any radiation from the system, required each signal delivered to a subscriber to be not less than 100 nor more than 3,000 microvolts, and specified that the bandwidth of each channel be at least 3.75 MHz. Beyond those three specifications other requirements were purely subjective and hardly constituted a technical standard. Amplitude linearity was specified as "The system shall be free of amplitude irregularities, i.e. system gains shall be constant over the working range of input signals to the system receiving antenna," and system degradation was specified as "The system shall be free of irregularities such as standing waves, spurious reception, etc. which result in an overall degradation of the redistributed picture." However, it did say that the receiving antennas of a system should be located within ten miles of the area to be served, and for some time this was used as a rule which inhibited the use of microwaves for program importation even from distant Canadian stations, a rule which was strongly supported by the CBC even though it was outside their jurisdiction.

This was the sum total of the regulations and performance specifications for CATV systems at that time, and remained so until 1968 when a new Broadcasting Act set up the Canadian Radio-Television Commission (CRTC) to regulate both broadcasting and cable TV, and a newly constituted Department of Communications (DOC) became the licencing authority for the technical aspects of the industry. Later, in 1971, the DOC issued a revised and more comprehensive technical performance specification as Broadcast Procedure 23 which replaced RRS102.

Certainly neither the regulations nor the technical performance standards of those days were comprehensive or unduly restrictive, and they did reflect the newness and almost total lack of technical sophistication of this still very young industry. This was recognized in the preamble to RRS102 which said, "The Department may revise this Specification as the state of the art advances or as requirements necessitate." It is significant that the only piece of test equipment referred to in the Specification is a calibrated microvoltmeter used for measuring signal output and radiation levels. All other characteristics of the system were to be tested using a test receiver viewing a broadcast test pattern, a subjective assessment.

The characteristics of a suitable test receiver for this purpose were specified in RSS102 as follows: "(a) bandwidth to enable a horizontal resolution of at least 325 lines; (b) amplitude linearity which will enable uniform resolution of each shade of the test pattern grey scale; and (c) distortion such that the reproduced picture is free of visible 'black-after-white' and 'white-after-black' trailing effects." Nothing very sophisticated or precise in that, the main item of test equipment to be used in the maintenance of a system to specified standards! This in effect was the extent of government regulation of CATV in Canada from 1956 up to the time when it was included as part of broadcasting under the supervision of the CRTC in the Broadcasting Act of 1968.

The first *Annual Report of the Board of Broadcast Governors* (BBG), published in June 1960 after the completion of the new agency's first full year of operation, included some comments which were portents of problems to come, some of which are still with us. For example, under the heading "United States Stations Serving Markets in Canada" it said:

> "Canadian listeners and viewers in such centres as, for example, Toronto and Vancouver, can receive a more varied broadcasting service through the opportunity to 'tune in' stations broadcasting from the United States. These U.S. stations nominally serve substantial audiences on their own side of the border. However, the licencing by the Federal Communications Commission (FCC) of stations designed primarily to take advantage of nearby Canadian markets may make it difficult for the Board to implement the policy of providing a service that is basically Canadian in content and character.
>
> "An application has been approved by the FCC for the establishment of facilities at Pembina, North Dakota. The location of these facilities in relation to populations south and north of the international boundary, and the activities of the licenced company, indicate that the principal market for this station will be Winnipeg, and this station may commence operation in advance of the second Winnipeg station for which a licence was recommended by the Board in January, 1960. The operation of the Pembina station as a third English-language station serving the Winnipeg market is not consistent with the policy of the Board and cannot but have a detrimental effect on the operation of stations in Winnipeg and on the service provided by them."

This was virtually the first official statement of the problem of the impact of American stations and programs on the Canadian television scene, which was intensified by the proliferation of CATV but certainly not caused by it. In 1960 there was no cable in the Toronto area, none in Winnipeg, and very little in Vancouver relative to the size of the market, so that the BBG's concern was really with the extent to which Canadians living near the U.S. border were receiving these stations on their domestic antennas. Today this concern is still evident but is related more to the impact on the Canadian broadcasting system of the many satellite-distributed U.S. program services in competition with similar services originating in Canada.

Under the heading "Competing Systems" the Board said in this first report:

> "'Pay-as-you-see' television is the subject of an experiment in Etobicoke, a suburb of Toronto, by Famous Players Corporation. This particular system transmits the signals over cable — they are not broadcast. The system is operating, and estimates of the number of present subscribers receiving service run between 3,000 and 3,500. No information has been released by the company as to the profitability of the enterprise but it is a fact that new subscribers are being hooked up daily.
>
> "It is an experiment that is being watched closely by motion picture and television companies in the United States, and it has been the subject of many articles in trade magazines, and has occasioned visits by American executives to the transmitting centre in Etobicoke. Members of the Board of Broadcast Governors

> also have inspected the centre and had its operation explained to them. The company so far has not released any figures as to the collections from the coin boxes attached to the receiving sets. Similar enterprises in the United States failed, but conditions and methods of operation in Etobicoke are different, and because of these factors U.S. executives are keen to ascertain the financial results and the public acceptance in an urban dormitory suburb.
>
> "No federal licence is required at the present time for 'pay-as-you-see' television using distribution by cable. There is some dispute as to whether 'pay-as-you-see' television by cable entirely within a province can, or should, be brought under the jurisdiction of the federal authority as it does not use Herztian waves as set out in the Radio and Broadcasting Acts. This is a matter that could be referred to the law officers of the Department of Justice for an opinion."

Some twenty years later the situation had not changed. The government still could not decide, even after two public hearings by the CRTC, whether pay-TV, as it was then referred to, was a good thing or a bad thing for Canada and Canadian culture, who should operate it if it were permitted, and for whose benefit. The only thing that had changed was the fact that, whereas the Etibicoke system was a closed-circuit operation on cable with no broadcast element involved and clearly not within the CRTC mandate, by 1980 any pay-TV operation would probably be using satellites for primary distribution and then be piggy-backed onto regular cable TV systems, and could not then be introduced without the approval of the federal authority having regulatory jurisdiction over both these elements of the system.

The BBG had no mandate under the Broadcasting Act of 1958 to regulate CATV, which was not considered to be part of the Canadian broadcasting system at that time. However, the Board did make a passing reference to CATV in their first annual report. Under the heading "Competing Systems: Community Antenna Television Systems" it said:

> "The other main method of transmitting television programs other than 'pay-as-you-see' television, and standard television using Hertzian waves, is community antenna television, commonly referred to as CATV. A CATV station receives broadcasts of commercial television stations from a high, specially constructed antenna, beyond the purse of most individuals, and then disseminates them to the public by cable. The public usually pays a connection charge and a monthly rental for this service. Inasmuch as there is a broadcast receiving station involved the CATV operator must be licenced by the Department of Transport and pay a $25 annual licence fee.
>
> "There is no doubt of the jurisdiction of the Parliament of Canada to legislate concerning this type of operation. There are now some 200 CATV stations in Canada which do an off-air pick-up of television programs from one to six television stations and then pipe the programs to subscribers' TV sets via land line. The greatest development of this service has been in Quebec where there are more than 100

> CATV companies. Ontario and British Columbia have roughly 30 each, and there are smaller numbers in New Brunswick, Alberta, and Saskatchewan.
>
> "The Canadian Association of Broadcasters has advised the Board that it would not be averse to necessary amendments to parliamentary enactments to bring CATV stations under the jurisdiction of the Board of Broadcast Governors. The CATV stations do offer competition to commercial stations in the fight for audiences, and they can pick up and transmit the programs of regular television stations without the permission of the originators or without any fee to such stations. The Board has had consultations with the Department of Transport and with the Canadian Association of Broadcasters on this jurisdictional issue and plans further meetings with officers of these two groups as well as officials of the CBC to arrive at a possible recommendation to Parliament for suitable amendments to existing legislation."

There, spelled out in no uncertain terms, was the writing on the wall, and the strongest justification for the need of an industry association.

At this time, in 1960, the CATV industry in Canada, which was barely eight years old, was serving less than 100,000 homes in the entire country and already the broadcasters were complaining about possible competition and looking to legislation to protect them against it. Certainly the majority of CATV systems were improving the reception of

U.S. border stations, but in most communities these stations were already receivable on good domestic antennas so that competition for audiences by these stations was not significantly increased, and this was largely offset by the similar improvements in reception of Canadian stations in communities on the fringes of their coverage areas.

At least it can be said in hindsight that even in those early days the broadcasters could see the potential of cable television. This was certainly more than could be said of the telephone companies, and particularly of the largest of them, Bell, in whose operating area, Ontario and Quebec, the majority of the new systems were located. It is only in recent years, when the potential of cable television for services other than broadcast entertainment has become so apparent as to be the subject of regular public discussion in the media, that the telephone companies have made moves directed towards the ultimate goal of taking it over, expounding the "one wire theory" which says in effect that all communications services to the ultimate user should be provided over one wire, and that wire should be owned and operated by the telephone industry.

By early 1960 the NCATA of Canada had ninety-four members operating CATV systems serving approximately 60 percent of all CATV subscribers. The association had increased by nearly 50 percent since 1958, and in fact was to increase again by a similar percentage to 133 members by 1962. Furthermore, as the membership increased the association represented a steadily increasing percentage of the industry in terms of subscribers served, since practically all of the larger companies joined. Most of the hold-outs were smaller companies or those just starting in business whose subscriber totals were less than the average for the industry as a whole.

At this time the majority of the systems were in Quebec and Ontario, and NCATA was based in Montreal, also the site of the annual conventions so far. The 1961 convention was slated for Niagara Falls, and it was decided that from then on all future annual meetings would alternate between Quebec and Ontario. However, the number of systems in British Columbia was increasing and, since the majority were relatively small and it was too great a financial burden to travel to conventions in the east, it was felt that some special arrangements were needed for an annual meeting which could be conveniently attended by these members. So in April 1960 the association organized the first western convention at the Georgian Towers Hotel in Vancouver, just two weeks before the main meeting in Montreal. It was on a smaller scale, but the success of these separate meetings was well evidenced by the numbers at the two conventions the following year, when there were 156 at the national convention in Niagara Falls with eleven trade members participating in the trade show, and seventy-five at the western meeting with five trade members exhibiting.

Very early in the association's history, in fact at the board meeting following the second convention in May 1958, a committee of three board members was appointed to study all aspects of the General Radio Regulations as they applied to CATV systems and meet with representatives of the Department of Transport to draft regulations more suitable to the requirements and development of the CATV industry. This was the first venture by the association into the political arena, but for some time it was not a particularly active venture. This was by intent since early meetings with the DOT in mid-1958 gave the clear impression that they were not anxious to discuss any outstanding problems in view of the pending appointment of a new regulatory agency which would take over responsibility for CATV from the Department.

The new Board of Broadcast Governors was appointed in November 1958, and with this appointment the regulatory future of CATV appeared to be clarified. The BBG mandate covered broadcasting only and CATV remained under the licencing jurisdiction of the Department of Transport and was not at that time included as part of the broadcasting system.

Nevertheless, there were some uneasy stirrings on this subject. At BBG public hearings during 1959 there were various references to the threat CATV represented to the policy of encouraging Canadian programming and limiting U.S. content, although the Board made no specific proposals for regulation. It was clear that these references had led to the comments by the BBG in their first annual report concerning consultations with the DOT, the Canadian Association of Broadcasters (CAB), and the CBC to arrive at a possible recommendation to Parliament for amendments to existing legislation. It was agreed within NCATA that, since the new Broadcasting Act had left the licencing jurisdiction over CATV undisturbed, it would be well advised to take no positive action on this or similar matters of government policy at that time unless forced to do so by further developments.

Such developments were not long in coming. In April 1960, on the suggestion of the Minister of Transport, the BBG called together a committee consisting of representatives of the Canadian Association of Broadcasters, the Canadian Broadcasting Corporation, and the Department of Transport to "consider the effect of Wired Systems on broadcasting in Canada." The first meeting of this committee was held on June 13, 1960, and it was considered ominously significant that NCATA, as the official representative of the industry being discussed, was not invited to participate. The explanation given was that the BBG had no jurisdiction over CATV operators as licencees, but NCATA pointed out that this very explanation negated any reason for discussing a service which was not within the Board's mandate.

In September 1960 NCATA decided to prepare suitable educational material for political and government circles on what CATV was and what service it provided to the Canadian public, and to prepare a brief to be presented to the relevant government agency at the appropriate time. To assist in this endeavour the board of directors decided to retain legal counsel, and selected Willard Z. Estey, later Mr. Justice Estey of the Supreme Court of Canada. "Bud" Estey had the reputation of being one of the foremost communications and broadcast lawyers in Canada, and he worked closely with the association on legal and political problems for the next ten years or more.

In November 1960 the Board of Broadcast Governors invited NCATA to attend a meeting of the Wired Systems Committee at which there was a full and frank discussion of CATV in general and the many practical and constitutional problems involved in any attempt at regulation. The industry representatives gained the impression that the BBG was fully conscious of the practical difficulties involved, but nevertheless considered it one of their major responsibilities to ensure that Canadian broadcasting acquired and retained a strong Canadian flavour.

There were two further meetings of this committee in December 1960 and January 1961 in which NCATA participated, following which the BBG submitted a report to the Minister of National Revenue giving the views of the committee on proposals for regulation. There were many references in the first drafts which were unfavourable to CATV and which were deleted or modified in the final report as a result of NCATA participation. The BBG Chairman, Dr. Andrew Stewart, was informed that, although NCATA had been associated with the preparation of the report, the association did not agree with all the views expressed and would submit its own brief to the House of Commons Special Committee on Broadcasting, which would be responsible for the detailed examination of any proposed legislation.

Such a brief was prepared and submitted but was never considered by the Parliamentary Committee because legislation was not introduced. As a result of what appeared to be a generally favourable situation in relation to the BBG and the Parliamentary Committee, and the probability at that time of the maintenance of the status quo on regulation, NCATA's national convention in May 1961 agreed that an approach should be made to the Department of Transport for some clarification and amendment of the Radio Regulations as they applied to CATV.

The nature of this approach was very controversial within the association, as some views were expressed that the requirement for licencing of systems should be completely removed so long as individual TV sets in homes were unlicenced. However, it was decided that the approach should use maximum diplomacy without any resort, at least initially, to political pressure. A list of problems and anomalies arising from the Radio Regulations, together with suggestions for appropriate amendments, was then discussed with officials of the Department of Transport and with the deputy minister at meetings in early 1962.

Meanwhile, to add to the developing political uncertainties in CATV, the Public Utilities Commission in British Columbia had announced that it intended to regulate systems in that province. It followed this up with a request that all systems apply for a Certificate of Public Convenience and Necessity and instructed the Electrical Energy Department not to issue any new permits except to holders of such certificates. NCATA referred the matter to the provincial Attorney General, who gave it as his opinion that CATV systems were public utilities under the provincial act. It was later agreed between NCATA and the Public Utilities Commission (PUC) that this requirement would be waived for association members until the validity of the decision had been tested in the courts. Legal counsel then prepared test cases involving three different systems which, between them, covered most of the varying circumstances under which CATV was operating at that time.

There was considerable discussion arising from Bud Estey's opinion that once the issue of provincial regulation had been raised in the B.C. courts it would automatically come to the attention of all other jurisdictions, and it was likely that the industry as a whole would eventually be regulated by one jurisdiction or another. If this were to be the outcome it was agreed that federal regulation would be preferable to regulation by ten different provinces, a tenet which has been the basis of association policy ever since. As a result, it was proposed that NCATA forego direct action in the courts of British Columbia in favour of persuading the Department of Transport to seek a ruling from the Supreme Court of Canada on the jurisdictional issue. This would be based on the grounds that acts of a provincial government were threatening to prejudice the enjoyment of a CATV licence issued by the federal government, and that since the industry was regulated under the federal Radio Act any attempted provincial regulation was outside their jurisdiction.

In January 1963 the BBG had denied an application by CJAY-TV in Winnipeg for a re-broadcasting station in Brandon, Manitoba. Subsequently the DOT issued a licence for a

CATV system in that community, so CJAY-TV applied for a rehearing of their application. At a public hearing in March 1963 the Board considered this application and reaffirmed its denial, saying that it was not persuaded that the licencing of a CATV system in Brandon provided adequate grounds for a reversal of its decision. However, during the hearing the Board referred to its concern over the growth of wired systems and announced its intention to hold a public hearing in June 1963 to receive representations from broadcasters and other interested parties on the relation between the development of wired television systems and broadcasting, and on broadcasting policy in general.

Following this announcement NCATA had several meetings with the Minister of Transport, the Secretary of State to whom the BBG reported, and the chairman of the BBG, as political interest in CATV was clearly increasing. At these meetings it appeared to the association representatives that the BBG was primarily using "wired systems" as an excuse to air the whole field of broadcasting policy which the Board felt needed clarification from Parliament. NCATA argued that the BBG had no authority to hold a public hearing on wired systems since these were not under its jurisdiction, but nevertheless agreed under the circumstances to submit a written brief presenting the industry's point of view on regulation and the appropriate regulatory jurisdiction.

It was intended that this brief would not be supported by representations in person at the hearing since this would tend to acknowledge the BBG's jurisdiction. But this decision changed when the BBG issued a press release with their official announcement which stated that "because of the vital nature of this hearing in the whole area of serving the national purposes through the broadcast media, the Board is making arrangements to have the proceedings transmitted on both radio and television."

NCATA's brief was presented at the hearing by Bud Estey, supported by Fred Metcalf representing operators in English Canada, and Armand Rousseau representing operators in Quebec. The brief concentrated on the "pure CATV, extension of an antenna" concept of the business. However, it was known that the main concern of the broadcasters was that the rapid growth of CATV could encourage its use as a vehicle for pay-TV with its expected ability to compete for Canadian content and talent. Therefore, Trans Canada Telemeter presented a separate brief dealing with pay-TV by wire and the use of networks for closed circuit theatre presentations. Individual companies also presented three other briefs dealing with closed circuit programming of a non-pay variety on CATV — what today we call community programming.

There was no direct or public outcome of this hearing since the BBG had no jurisdiction to issue any decisions or conclusions on the subject. But it certainly brought the existence and growth of the CATV industry to the attention of the policy makers for the first time in the short history of the industry in Canada. During the succeeding months there were questions asked on the subject in the House of Commons, and there were a number of meetings between NCATA representatives and several cabinet ministers, individual Members of Parliament, and caucus committees of both the Liberal and Conservative parties, as well as approaches by various members of the association to their own MPs.

Two developments in particular were giving the government cause for concern at this time. One was the increasing penetration of the Canadian market by U.S. broadcasters near to the border which was blamed, at least indirectly, on the growth of CATV. The most blatant example of this had been the approval by the FCC of a licence for a high power broadcast transmitter at Pembina, North Dakota. There was no sizeable market in North Dakota for this station, and since the transmitter was to be located only three or four miles from the Manitoba border and little more than fifty miles from Winnipeg, it was clearly aimed at this Canadian market.

The second development was an option negotiated during 1963 by Columbia Broadcasting System, the U.S. television network, to acquire a 75 percent interest in Vancouver Cablevision Ltd., which included CBS putting up all the financing required for further expansion. At this point Famous Players Canadian Corp., a Canadian public company of long standing whose shares were listed on the Toronto Stock Exchange but 51 percent owned by Paramount Pictures based in New York, had been aggressively developing interests in CATV since 1960. This was undoubtedly causing some concern in government circles, particularly as Famous Players had been operating the Telemeter pay-TV experiment in Etobicoke since its commencement in 1960. However, the CBS purchase seemed to be the trigger which precipitated action.

On November 29, 1963, J. W. Pickersgill, the Secretary of State, made a statement in the House of Commons that the government was considering further regulation of CATV. On December 31 Pickersgill and the Minister of Transport issued a joint statement in effect imposing a freeze on the issuance of further CATV licences and referring the subject of regulation once again to the BBG for study and recommendation. The announcement did not refer to a licence freeze as such, but clearly this was its effect, since NCATA began receiving reports that DOT district offices had been turning down applications for licences, particularly where they included reception of U.S. stations, citing the discretionary powers of the minister as their authority.

Coincident with this announcement the ministers requested the Board of Broadcast Governors "to inquire into and recommend any legislative action that may be required to ensure that, so far as the constitutional jurisdiction of Parliament will permit, the use of community antenna television for the dissemination of television programs is subject to similar regulation under parallel conditions to that applied to direct broadcasting." To implement this request the Board appointed a joint committee of the BBG and the DOT, and on January 9, 1964, NCATA representatives met with this committee and presented the views of the CATV industry.

These views were summed up in the words of a brief submitted by NCATA representatives to the committee:

> "The Government seems to have already assumed that cable TV is adversely affecting the national purposes, but does not seem to have realized that the 5 percent of Canadians using CATV whose television viewing it is proposed to regulate hardly constitutes a large enough group to be of great concern. The fact that 50 percent of Canadians can view American TV off their home antennas makes it obvious that the CATV viewers are of no real significance in the 'national picture'. Indeed, in many areas if CATV is regulated as to American content the Canadian viewer will get around this regulation by simply re-erecting the house antenna which he took down a few years ago upon the advent of CATV in his community. This retrograde step will be forced on him by government censorship over CATV which does not apply to individually owned antennas. It appears to us that it is not the prerogative of the

> government to force the Canadian public to watch Canadian programs by prohibiting all the alternative programs in the air. The mere fact that regulations prohibit some Canadians from watching American programs will not ensure that they will watch Canadian programs anyway."

This brief clearly summarized the problem which has always faced broadcasting in Canada due to the intrusion of stations in border states of the U.S.A., whether these are located intentionally such that they cover homes in adjacent Canadian territory or not. This is still, more than thirty years later, a basic problem of Canadian broadcasting — indeed even more so now in the age of satellites which, to an even greater extent, recognize no international boundaries.

The committee report was submitted to the ministers and tabled in the House of Commons on March 19, 1964, with recommendations that effective control over CATV could be achieved by amendments to the Broadcasting Act which should include:

(1) a requirement that the Minister of Transport refer any applications for licences, including renewal of existing licences, to the BBG for hearing and recommendation and give the Board the necessary powers for this purpose;

(2) authority for the Board to regulate such stations in regard to program content and other non-technical matters, (the Department of Transport would continue to deal with technical aspects);

(3) extension of the objectives and purposes of the Broadcasting Act to Commercial Broadcasting Receiving Stations and to Land Stations feeding cable relay distribution systems ("Commercial Broadcasting Receiving Station" and "Land Station" are terms used by

the Department of Transport for the antenna systems and other receiving equipment used to feed off-air signals to CATV systems);
(4) a limit on foreign ownership of new Commercial Broadcasting Receiving Stations and Land Stations feeding cable relay systems similar to the limitations contained in the Act for broadcasting stations;
(5) authority for the Board to control interconnections (networks) of cable relay distribution systems when they are fed with signals from Commercial Broadcasting Receiving Stations or Land Stations;
(6) authority for the Board to enforce of its own action conditions of licences of stations in these categories where the conditions have been included as a result of recommendations from the Board, and to take such action in the courts on violations of such conditions or of infractions of the regulations.

The Broadcasting Act was not in fact amended at this time as recommended. Instead on July 22, 1964, the Minister of Transport announced that, while CATV systems would remain under his licencing authority, all applications would be referred on an informal basis to the BBG for recommendation. In his statement he said that "the so-called freeze will be lifted immediately and the Minister will deal with pending applications after informal consultations with the Board of Broadcast Governors on the possible effect on television broadcasting in the areas concerned."

This was not a very satisfactory situation for the CATV industry. Applications were considered in camera by the BBG, and there was no procedure which allowed for any representations concerning an application to be made directly to the Board. In fact, during the eight months after this statement was made, up to March 31, 1965, the Board examined fifty-nine applications and recommended against thirteen, or 22 percent, of them. The BBG was unhappy with this piecemeal approach and said as much in their annual report of March 31, 1965. Nevertheless, this ad hoc arrangement continued for four years until the passage of the new Broadcasting Act in 1968, which established the Canadian Radio-Television Commission on April 1 in place of the Board of Broadcast Governors, and defined cable television as an integral part of the Canadian broadcasting system, bringing it by law under the regulatory control of the CRTC.

The hectic NCATA legal and political activity of the latter part of 1962 and the first part of 1963 brought to a head the difficulty of running a growing association on a voluntary part-time basis, as it had been since its inception in 1957. As early as the third convention in May 1958 the president at that time, Armand Rousseau, had forecast the need for a full-time secretariat as the association grew larger with the development of the industry, and had pointed out the need for a considerable increase in financial resources to cover this contingency when it arose.

By early 1963 NCATA had 145 active members and nine trade members, and yet the total annual income was barely $10,000. In fact the association had an operating deficit for the first time due to legal expenses not incurred before. Not only was it necessary to budget for substantially increased legal expenses and public relations costs during the next year, but it finally became clear to the board of directors that major changes in the administration were inevitable in order to cope with the regulatory and other problems which were then pressing on the industry, and which were the main reason for the organization. Accordingly a budget of $60,000 was adopted for 1963-1964, which included approximately $30,000 for an office and full-time secretariat. The dues structure of the association was changed to one based on the number of subscribers, and a search committee was appointed to consider possible candidates for a full-time general secretary.

During this search process it was learned that there were several companies which specialize in professional association management. These companies provide all administrative, accounting, public relations, and similar services required by trade associations on a multiple basis. A company might have as many as six or seven associations as clients, so that the basic overhead costs are shared between them. After discussions with four such management companies it became clear that this was not only a practical solution to the association's administration problems, but would actually cost less than the alternative of employing a manager and support facilities on a full-time basis.

Accordingly in November 1963 Chevalier & Associates was hired to provide these services. This selection was determined partly by the fact that the company was located in Montreal where the association was based, while the other three were in Toronto, and partly by its bilingual ability which met the need to be able to operate equally in English and French, since almost half of the members at that time were located in the province of Quebec. While elected officers and directors of the association continued to handle the regulatory problems and relations with public utilities, etc., Jacques Chevalier, one of the company principals, was appointed manager, and he and his staff took over many of the administrative responsibilities, including accounting, dues billing, membership correspondence, and organization of the annual conventions.

Experience proved that professional association management was an excellent arrangement, but in a trade association with a rapidly growing membership and increasing involvement in the political arena it had its inevitable limitations, and by early 1968 these were being increasingly felt. The basis of this arrangement was the availability of professional management facilities coupled with the economy of sharing the overhead costs with other client associations, but eventually this became a constraint as the increasing workload strained these shared facilities.

These constraints became particularly evident during 1967. The pressures of the evolving political situation were building up to their final conclusion in the passage of the new Broadcasting Act in March 1968, which created the Canadian Radio-Television Commission and designated cable television as part of the broadcasting system under the regulation of the Commission. Even though all the daily administrative details no longer required the time and attention of the officers and directors, the pace of regulatory developments demanded more time than could be reasonably given by part-time volunteers. It was this rapidly changing environment that persuaded the association to change its name to the Canadian Cable Television Association or CCTA, thus formalizing references to the industry from "CATV" to the more descriptive "cable television."

In May 1967 John Loader had been elected president of the association, and at the end of his term the following year, since he was in semi-retirement, he was persuaded to become executive vice president on a part-time basis. He provided very valuable service to the association and to the cable industry in this capacity, especially in helping to establish a good working relationship with the newly formed CRTC through the early years in 1968 and 1969 and this involved him in frequent commuting between his home in Victoria and Montreal and Ottawa. He gave unstintingly of his time, energy, and ability, although he was in advancing years. In fact this constant untiring devotion to the affairs of the association and the industry it represented cost him his life. In February 1971 while attending a seminar at the University of Toronto on "Communications and the Public Interest" accompanied by the association's founding secretary, Ken Easton, Loader suffered a heart attack from which he died two days later.

By mid-1969 it was evident that it was necessary to establish a separate office with full-time personnel. This was confirmed and recommended by a management consulting firm retained to examine the organization and policies of the association. As a result in October 1970 the board of directors announced the appointment of Bob Short as the first full-time president to begin his duties on January 1, 1971, with the executive office located in Ottawa.

Since that time the cable TV industry has grown steadily in both Canada and the United States, and with it the national associations representing the industry in each country. In Canada, with a total of 9.8 million television households, all but 100,000 are in areas which are licenced for cable TV and of these 98.6 percent are passed by cable and therefore have access to cable television services. More than 80 percent are connected and receiving service from almost 2,000 individual systems. In the United States there are 92 million TV households and 96 percent of these are passed by cable. Sixty-two percent of these are connected and receiving cable TV service from nearly 11,000 individual systems.

The growth in cable penetration, that is the percentage of the homes passed by cable which are connected as subscribers, during the last decade has been more marked in the U.S. than in Canada. In 1980 Canadian penetration was 57 percent compared with the current 80 percent, while in the

United States it was only 22.6 percent compared with the current 62 percent. This was due to the fact that cable TV was an established service in all the major centres in Canada by the 1980s, while development of most of the major metropolitan areas in the U.S. did not start until around 1980 after the introduction of national TV distribution by satellite in 1975, which effectively bypassed the FCC ban on reception for cable distribution of "distant stations."

However, the growth in pay-TV has been much greater in the United States in part because the situation was reversed. Pay-TV service was offered throughout the U.S.A. from 1975 when Home Box Office went to satellite, and 77 percent of cable TV subscribers now subscribe to these services. In fact this level of penetration had been achieved by 1983, after only eight years. and has been almost constant since then. In Canada pay-TV was not approved by the CRTC until 1982 and not in effective operation until 1984, and today only 12.2 percent of cable subscribers use this service.

During the first decade of its existence after inauguration in 1957 the Canadian Cable Television Association (CCTA) had to devote practically the whole of its limited resources and manpower to the developing political and regulatory situation. After the establishment of the CRTC in 1968, and the inclusion by the new Broadcasting Act of cable as part of the broadcasting system, the next few years required a concentration of effort on acclimatizing the industry to the new regulatory environment and developing a firm working relationship with the new Commission.

During this period the cable industry grew steadily in size and with it the membership, and consequently the resources, of the association. In 1968 there was a total of 377 licenced cable systems in Canada, and these systems had cable in place capable of providing television service to less than 30 percent of the total homes in Canada. Of the homes to which cable was available about 43 percent were connected for service, so that of all the homes in Canada about 13 percent were cable subscribers. By 1978 the number of homes passed by cable had grown to 75 percent of the total homes, and 68 percent of these were subscribers, so that by this time more than 50 percent of all homes in Canada were obtaining television service from cable.

The industry, and with it the strength of the association, continued to grow during the 1980s. By the beginning of the 1990s the substantial increase in income due to this growth, coupled with a full-time professional staff, permitted an expansion of services to its members and this is reflected in several initiatives which the CCTA introduced or sponsored in the early 1990s.

In 1979 the CBC had been licenced by the CRTC to televise live Parliamentary debates from the House of Commons — the first live television production of a legislature in the world. In 1986 the cable TV industry proposed to enhance this service to include additional public affairs programming and, in cooperation with the CBC, formed the Canadian Parliamentary Channel, which was approved by the House of Commons with all-party support. However, in 1991 the CBC announced that it would cease its funding because of budget cut-backs, and the House of Commons itself began to pay for satellite transmission of its proceedings at an annual cost to the taxpayers of $2 million. Then in 1992 a consortium of thirty-three Canadian cable companies, christened the Cable Public Affairs Channel (CPAC), was formed to provide this service at no cost to the taxpayers or the cable subscribers.

Since then CPAC service has been expanded to include, in addition to full unedited coverage of all sittings of the House of Commons, broadcasts of major speeches, conferences, conventions, the proceedings of the Supreme Court, Royal Commissions, etc. Its programming schedule includes an average of 140 hours per week, and it is distributed nationally by satellite to cable systems across Canada serving 7.2 million homes at an annual cost, absorbed by the cable industry, of $3.1 million. From two staff members in 1992 CPAC has expanded to a full-time staff of thirty-five, augmented by various freelancers across the country.

In 1995 the cable industry introduced "Cable In the Classroom" to the educational community, coordinated by the CCTA. This industry initiative is committed to the distribution of educationally relevant, copyright-cleared, commercial-free, French- and English- language cable TV programming services to publicly funded schools across Canada. Participating cable operators agree to install and provide ongoing cable service, free of charge, to elementary and secondary schools passed by cable within their licenced service area, and to work with educators to utilize Cable in the Classroom to the maximum advantage. These efforts at the local level are supported at the national level by staff developing materials to support cable companies as they work with local educators, providing a regular schedule of Cable in the Classroom programming, and generally promoting media literacy. Educators are able to tape programs of interest which have been pre-cleared for copyright and are without commercial interruption, and replay them in the classroom as a means of enriching instruction.

In 1997 the Canadian Cable Television Association celebrated its fortieth anniversary. This marked a milestone in the growth of both the cable TV industry in Canada, and the association representing it. However, it also began to reveal problems the industry was facing due to the changing climate in which it was operating. The rapidly evolving technology, bringing with it government-sanctioned competition, had completely changed the operating environment, presenting new challenges for the individual cable companies and for the association representing them.

In 1996 the CCTA had received a number of applications for membership from competing distributors of broadcast services and from cable companies owned by competitors to the cable industry. The association had to deal with the situation of a proposed merger between a major telephone company and a mid-sized member cable company, and other members had received licences to operate services competitive to cable and had begun operations. Within the cable industry itself consolidation continued at a rapid pace, so that by 1997 the four largest companies were serving more than 75 percent of the total cable subscribers in Canada. While the CCTA represented some seventy other companies, there were more than one hundred smaller companies that were not CCTA members, and three of the larger companies had resigned their membership within the previous eighteen months.

It was becoming clear that with the arrival of competition and with the prospect of convergence blurring the lines between cable TV and other telecommunications services, the whole basis of CCTA representation of the industry and the relationship of the association with other similar organizations required a detailed review. In late 1997 at the request of the board of directors, Pierre Simon, a past chairman of the board, undertook such a review. This review included consultations with many members of the association and resulted in a number of recommendations to the board concerning the role and future actions of the CCTA.

It was clear from these consultations that many members felt strongly that the CCTA should continue to focus on serving and promoting the interests of the cable TV industry, and that, while some members may well be involved in other businesses, even competing ones, the interests of traditional cable television must always come first.

At the same time it was recognized that, once the cable industry is operating in a fully competitive environment and the regulatory regime is not only determined but fully implemented, the role of the association could change. It was therefore recommended that the mandate of the CCTA should be to continue to represent the interests of those companies who derive the majority of their revenues from the distribution of broadcasting services over a terrestrial network. However, within this constraint the association should devote more resources to communications with members, and to providing assistance in particular to smaller members in dealing with legal and technical problems arising from competition and changes in the marketplace.

Chapter 8

From cable TV to competition and convergence

From 1971, when the Canadian Cable Television Association was reorganized to handle the increasing workload and moved to Ottawa with a full-time president, the growth of the cable industry in Canada and the association representing it were both tied closely to government policies as defined by the regulations of the Canadian Radio-Television Commission, which by this time was well into its regulatory stride.

When the new Broadcasting Act came into force on April 1, 1968, the CRTC became the licencing authority for all broadcasting undertakings, including CATV systems, and all holders of CATV licences issued by the previous authority, the Department of Transport, were advised that they would have to apply to the Commission for licences under the new Act. This presented the Commission with an immediate heavy workload since there were at that time 377 CATV systems licenced by DOT, all of which were required to be dealt with individually under the act through a public hearing process.

As a result the CRTC, in its first year and a half of operation, concentrated entirely on existing DOT licencees and did not accept applications for new licences until late in 1969.

During this period the priority necessarily had to be to bring all current licences under the control of the Commission without waiting for a regulatory framework for the orderly development of cable systems as an integral part of the national broadcasting system. Policies and procedures covering such things as ownership, rates, areas of service, access for local and educational services, and relationships with broadcasters were built up partly as a result of these initial public hearings. Finally, late in 1969, the CRTC published a rewiew of its proposed regulatory framework for "distribution undertakings" — the term it had adopted for cable TV systems — and in July 1971 issued detailed regulations.

The intent of these regulations, as spelled out in a preamble, was to "clarify and strengthen the basic broadcasting services to Canadians, to increase the diversity of program services available to Canadians, and strengthen the means by which the whole broadcasting system will prosper, ensuring that one element of the system shall not prosper at the expense of another. The Commission's objective is to encourage the system to provide the best possible service and widest choice from every source for the Canadian viewer, while favouring services that are as relevant as possible to the particular community, local, regional, or national, that is to be served. The policy is intended to reduce any harmful impact of cable systems on local broadcasters while, at the same time, assuring the continued growth of cable television."

The policy was to be implemented by several mandatory requirements which formed a part of the conditions of each system licence. Cable systems were required to carry stations in a priority order starting with all "local stations," followed by all Canadian stations whose Grade B contours enclosed any part of the licenced area, followed by any distant Canadian stations not affiliated with any of the higher priority stations. Once all these priorities had been accommodated on a system, "optional" stations could be carried, including U.S. commercial and non-commercial stations.

The CRTC also encouraged local program production, although this was not made mandatory and, reversing a long-standing policy of the regulatory agencies, was prepared to authorize the use of microwaves for carriage of distant or optional stations. However, the number of channels carrying signals from U.S. commercial stations was generally limited to three. The Commission was also prepared to authorize the linking, or networking, of cable systems for the purpose of sharing a common headend or microwave transmission from a distant headend. The CRTC also recognized the importance of establishing long-range policies for the broadcasting system, and that the financial stability of a cable system could be influenced by the length of the licence term granted. It therefore announced its intention to grant licences for five-year terms, the maximum permitted under the Broadcasting Act, rather than the one- or two-year terms which had been prevalent.

Perhaps the most significant part of the new CRTC policies dealt with the relationship between cable television and broadcasters. In the past it had always been a basic requirement of licencing policy that stations received off-air for distribution by cable should not have their program content altered nor any part of their program or commercial content deleted. It had also been a dogma of the cable industry that cable distribution is nothing more than a passive antenna and systems should not be required to pay for that which is broadcast for any receiver to pick up. However policies and dogmas must change with the times.

The Commission was very concerned that some $12 to $15 million a year of Canadian money was being spent to buy commercial time on U.S. border stations whose signals spilled over into Canada. They intended that as much as possible should be repatriated and hopefully added to Canadian broadcasting revenues. The subject was extensively discussed at a public CRTC "Policy Hearing on Cable Television" in April 1971 at which many adverse views were expressed by broadcasting interests, which collectively implied that further development of cable in Canada would kill Canadian broadcasting.

This opposition was countered by a proposal from the CCTA that where a program was being aired on a channel received from a U.S. broadcasting station and simultaneously aired by a Canadian station carried on another channel of the cable system, the Canadian version with its accompanying commercials would be carried on both channels. This was termed "simultaneous substitution" and was accepted by the Commission in policy statements resulting from the hearing. For this purpose the CRTC withdrew the requirement that received signals could not be altered, and permitted the removal of commercials from U.S. stations and their replacement by commercials sold by the Canadian stations. In addition, cable systems carrying U.S. stations were required if requested by a Canadian broadcaster to carry its version of any simultaneously duplicated program.

The import of these new policies was far-reaching. There is no doubt that they confirmed cable TV as a permanent and important member of the broadcasting community and afforded it the respect that had always been lacking. No longer was it the problem child that everyone tries to ignore and wishes would just go away. Incidentally the stature of the Canadian Cable Television Association increased in the eyes of the CRTC and others since many of the policies adopted had been recommended or supported by the association during the long consultative process leading up to the new regulations.

The CRTC had virtually freed cable systems from any regulatory restraints on the distant signals they could carry, particularly from the U.S. They had an obligation to carry a basic Canadian service, but once the Canadian priorities were satisfied there was no limit imposed on additional Canadian or U.S. stations, even to the extent that microwave was now permitted for this purpose. Also, under the new rules local program production was to be encouraged but not legislated. This was important because it gave systems the freedom to experiment and innovate without the constant threat of regulatory pressure, thus allowing this new type of service to find its own level in terms of acceptable form and cost, while not saddling the less profitable systems with substantial expenditures which could not be met without additional revenue.

In retrospect this major change in the regulatory environment had a beneficial influence on the developing cable television industry. True, it had imposed certain limits and conditions which did not exist before; it had increased the cost of operating a cable system without a commensurate increase in revenue, and it had included some unpopular decisions. Nevertheless, the industry had acquired a status which it did not have prior to 1968 and could now consider itself an equal in the Canadian communications scene with broadcasters and telephone companies. It was difficult for many people, particularly those who are instinctively entrepreneurs, to admit that industry growth could take place because of regulation; however, without the regulated order that the CRTC brought to cable television, the industry's development may soon have been showing signs of being stunted by unregulated competition, seriously eroded standards of performance and service, and a shortage of capital due to lack of interest by serious investors and financial institutions.

These beneficial effects are well illustrated by the statistics on the growth of cable TV during the ten years following the Broadcasting Act of 1968. *A Special Report on Broadcasting in Canada 1968-1978*, published by the CRTC in 1979, showed that in 1968 there were 445 systems licenced by the CRTC, and by 1978 this number had increased by 18.4 percent to 527; however, the percentage of total homes passed by cable had increased from 30 percent in 1968 to 75 percent in 1978, while the homes subscribing to cable had risen from 13 percent to 52 percent of the total households. It is evident that the dramatic growth in cable penetration during this period had been generated by the consolidation and expansion of existing systems and by attracting new subscribers rather than by growth in the number of systems.

This growth has continued to the present day. By 1998 the number of licensees had increased to 2,089, serving a total of 7.9 million subscribers and representing 91 percent penetration. Much of this increase in the number of systems can be attributed to the many small communities, which in earlier days had been too remote from broadcasting facilities or too small to justify the cost of a cable system but could now access television services delivered by satellites.

One feature of the new broadcasting regulations which had an immediate and visible effect on the development of cable was the willingness of the CRTC to permit the use of microwave for the carriage of distant or optional stations. This was a reversal of the policy which had existed since the earliest days when the licencing agency required the headend to be situated within ten miles of the area to be served, and the use of microwave, even for this limited application, was not permitted.

Basically there were two different applications of microwave to cable systems. One was to to replace cable as the primary link from the headend to hubs in a large system, or to enable the headend to feed an adjacent area where the service is required. In this application the microwave link, which is invariably one hop (a transmitter feeding one distant receiver), is really an extension of the trunk cable carrying everything being carried by the cable and must therefore have a multi-channel capability equivalent to that of the distribution system of which it forms an integral part. The other was to access a relatively remote headend location where improved off-air reception was available but the distance involved made cable access either impractical or too costly.

The first application required the use of multi-channel microwave facilities, since the system would be replacing a portion of the cable and would need to carry the full complement of channels; however, there were at that time no microwave frequency allocations in Canada suitable for this purpose. In the United States the FCC had set aside a portion of the spectrum from 12.7 to 12.95 GHz into which all requirements for microwave in association with cable TV were allocated. This was known as the Community Antenna Relay Services (CARS) band and was wide enough to accomodate up to forty TV channels.

In 1972 the federal Department of Communications (DOC) announced their intention of allocating a band from 14.4 to 15.35 GHz for cable TV use in Canada. However, the Canadian Cable Television Association opposed this on the grounds that it was impractical since there was no equipment available for this band and the Canadian market was not large enough to recover the costs of development. Eventually in 1974 DOC announced their approval of a microwave band equivalent to the U.S. CARS band specifically for use on single-hop intra-city systems requiring four or more channels, but not intended for multi-hop long-haul networks. This satisfied the needs of the first application and, starting in 1976, a number of systems of this type were installed in various cable systems in Canada.

Although used primarily to improve the performance of the primary link in the distribution system by replacing cable with lower loss microwave, in many of these applications the output from a multi-channel transmitter located at the head-end was split and beamed to several receivers at strategically located hubs within the distribution system. In this manner a headend could serve several cable systems located within microwave range, or a large metropolitan area could be broken up into smaller discrete areas for distribution purposes — in fact the beginnings of the "hub" type of distribution which is typical of most large cable systems today and has facilitated the current application of optical fibre transmission.

The second application of microwaves to cable TV, access to distant headend sites, could not be satisfied by the use of this band because, although only one to four channels were generally needed for this purpose, the greater distances involved required multi-hop facilities, several hops in cascade. Usually these could only be provided by leasing from common carriers, typically the telephone companies. With this easing of regulatory prohibitions major use of microwave, both leased and privately owned, developed across the country. By 1980 there were leased long-haul systems in operation in every province of Canada, each carrying two, three, or four channels of programs originating from U.S. stations received close to the border and providing U.S. network service to more than fifty cable systems, large and small, to whom this service was not available by direct reception.

The longest of these was a system originating at a head-end in Chamcook, New Brunswick, where signals could be received from U.S. network outlets in Maine and serving cable systems in all four Atlantic provinces. The microwave system, leased from the telephone companies in each province, extended for more than 2,000 km (1,200 miles) through New Brunswick, Nova Scotia, Prince Edward Island, and Newfoundland to St. John's, providing service on four channels to twenty systems serving nearly 200,000 subscribers.

Similar microwave systems existed in every other province, each having a capacity of two to four channels carrying U.S. network services to systems throughout the province being served; however, these were not inexpensive facilities. The total annual cost of these was some $6.5 million, but they were providing U.S. network service not otherwise available to 1.2 million subscribers at an average annual cost of $5.60 per subscriber, or forty-five cents per month.

By 1980 satellite distribution of television services had been proven feasible and was becoming relatively commonplace in the United States. A communications satellite is in effect a microwave repeater station in space capable of receiving signals in one frequency band from a powerful transmitter directed towards it from earth, converting them into another frequency band, and transmitting them back towards the earth. The "downlink," as it is termed, has a fairly broad beam designed, for example, to cover the entire country, so that signals can be received with a suitable antenna anywhere within this coverage area.

In order to achieve this designed coverage and to be capable of being received by an earth station without the need for complicated tracking facilities, a satellite has to be located 35,800 km (22,250 miles) above the equator in what is known as the geosynchronous orbit. In this position it orbits the earth at the same speed as the earth revolves so that the satellite appears to be stationary in relation to the earth's surface. A number of satellites designed to uplink in the 2 GHz band and downlink in the 4 GHz band can be accommodated in this orbital arc, providing they are separated along the arc by at least three to five degrees of longitude so that an earth receiving station, using its directional capability, can receive the signals from any one satellite without interference from any of the others.

The first application of geostationary satellites to communications was for international telephone service. One of the earliest was the Canadian satellite Telstar, which provided telephone service across the Atlantic and promised a cost effective and technically superior alternative to the existing submarine cables. Coast-to-coast television by satellite was first demonstrated in June 1973 at a National Cable Television Association convention in Anaheim, California, with transmission of an opening address from Washington D.C. Following this demonstration, in September 1975 Home Box Office commenced distribution of their pay-TV program service nationally using a transponder leased from RCA on one of their early satellites.

It was a tremendous and courageous gamble for Home Box Office because they signed a six-year contract with RCA without having contracts for more than two or three potential users to support it, and without even the assurance that it would work technically. No one could be sure at that time that the use of lower cost cable-owned-and-operated earth stations would be satisfactory, but HBO could see the tremendous potential if success could be proved. However, it did work and within little more than two years there were some

150 earth stations in the United States specifically serving cable TV systems, and the number of subscribers using the HBO pay-TV service had increased to more than 800,000 in some 375 systems, literally from one end of the country to the other.

Canada was the first country in the world to launch a domestic satellite intended for communications within the country, as distinct from the Intelsat series which were designed for communications between continents. To finance and operate this new form of communication the government in 1969 incorporated Telesat Canada as the nation's domestic satellite communication carrier. Telesat not only owned and operated the satellites but was the only licencee approved to own and operate both the transmitting and receiving earth stations. Telesat was incorporated with half the issued shares owned by the federal government and the other half owned by Trans Canada Telephone System (TCTS), a consortium of the nine major telephone companies operating in the ten provinces. The government promised that once Telesat was in successful operation, half the shares would be made available for public subscription.

This never happened. Instead in 1976, following intense lobbying by the telephone companies, the federal cabinet authorized what amounted to a merger between Telesat Canada and the Trans Canada Telephone System against the advice and recommendation of the CRTC which, under new legislation that year and a new name (Canadian Radio-television and Telecommunications Commission), had become the licencing authority charged with regulation of federally regulated carriers. Under the influence of the telephone companies Telesat proceeded to set tariffs for these services which were clearly not attractive to private users, and appeared to be designed to protect the terrestrial microwave services offered at similarly inflated rates by those same carriers.

This situation seriously restricted the development of Canadian domestic satellite services and allowed the United States, which from the start had adopted a policy of free enterprise competition, to pull way ahead, although they did not launch their first domestic satellite until some two years after the first Canadian Anik was flying. By 1979 there were four Anik satellites in orbit with a total capacity of at least forty transponders, yet they were grossly underutilized largely because of cost. Only three of these transponders were in use for video services, and the sole user of these was the CBC. By comparison there was a total of seven American domestic satellites in orbit carrying between them twenty-one operational video services, the majority of which were intended specifically for cable TV reception and distribution.

The federal government came under considerable pressure to break this deadlock by permitting private ownership of Television Receive Only (TVRO) earth stations, since this would at least help to reduce some of the excessive costs involved. Furthermore, if the use of earth stations proliferated then it might be expected that this would provide the incentive for increased use of the satellites, which should in turn permit a reduction of the transponder charges.

Finally in February 1979 the government allowed private ownership of earth stations. This did not, however, lead to a rapid proliferation as might have been expected for two reasons. First, licenced stations were confined by the terms of their licences to receiving only Canadian program material, and the only such material distributed by satellite at that time was the proceedings of the House of Commons produced by the CBC. Second, all satellites then operating, both Canadian and American, transmitted to earth in the 4 GHz frequency band, a microwave band shared and heavily used by common carrier terrestrial systems.

The problem this created arose from the fact that the signal received from a satellite is very weak — after all, the transmitter is 35,800 km away and by any standard that is a long microwave hop. For this reason parabolic antennas at least twelve to fifteen feet (3.5 to 4.5 meters) in diameter were required in conjunction with special low-noise amplifiers. Without these devices the signals would be virtually unusable; however, such an installation was equally sensitive to other signals in the same band, and that meant signals being transmitted by terrestrial microwave stations. Whether a microwave transmitter will cause interference with satellite reception at a TVRO earth station operating in the same 4 GHz band depends on many things: the distance from the TVRO site; the ground profile between them; the amount of power it is radiating and its direction from the TVRO site; the direction in which the microwave transmit antenna is beamed towards its own receiver; the characteristics of the antennas; etc.

This extreme sensitivity to interference from terrestrial microwave stations meant that the choice of a suitable location for a TVRO earth station involved entirely different considerations from those which applied to the selection of a cable TV headend site. A headend requires maximum elevation in order to receive distant stations with the strongest possible signal, while elevation is of no importance to an earth station which is looking up at the sky, but protection from earthbound signals in the same frequency band is essential. It was said that the best location for a headend is the top of a hill while the best location for a TVRO earth station is a hole in the ground — and this was almost literally true!

Because of the amount and nature of the information that was required to assess the suitability of a proposed TVRO site, and the number of potentially interfering microwave stations that could be involved, assurance that there would be no interference with satellite reception could only be given after a rigorous analysis based on access to a computer database. All this detailed information had to be included in the form of an engineering brief which formed part of an application for a licence, and this added to both the cost and the time required to establish a TVRO earth station and necessarily limited the number which were built.

Since reception from satellites is available virtually anywhere within the coverage area of the satellite transmitter independent of distance or location, one result of their use for national television distribution — which was perhaps unexpected and certainly not planned for — was the ability of homes in remote areas far from any possible off-air reception of U.S. program services to obtain these services by reception from the satellites which were now carrying them. True, this reception involved the use of an antenna which, while not as large as those required for TVRO stations, was still large enough to be as expensive as a domestic installation.

Neverthless, by 1978, only three years after the start of the satellite TV era, there were no less than twelve transponders on American satellites carrying full-time TV services. These were sufficient to encourage many households, particularly in rural areas too sparsely populated to support a cable system, to invest in satellite dishes. This was really the beginning of direct-to-home television, or DTH as it became known, and the first potential competitor to cable as an alternative to off-air reception of broadcast television.

In Canada by this time some 75 percent of all homes had access to cable TV and just about every city and town of any size was wired, but cable TV had never been considered a practical service for rural or remote areas. The population of Canada has always been concentrated in a band along the southern border, leaving immense areas remote from the facilities available in the south, including broadcast television. Rural and remote areas have two characteristics which handicap them for television reception. First, because economics dictate that broadcasting stations cover as large a population as possible, it is inevitable that these stations will be located where they can cover populous areas, and that means in the vicinity of cities or groups of towns. Second, since the homes in these remote areas are not crowded together as they are in urban or suburban areas, if cable were installed the density in terms of homes per mile of cable would be too low to justify the cost to each home. Direct-to-home broadcasting by satellite appeared to be the obvious answer to this problem, but the cost of a satellite dish at every home made it prohibitive except to the fortunate few who could afford one.

In 1977 a group based in the Yukon and headed by Rolf Hougen, a businessman and broadcaster in the territorial capital Whitehorse, presented a proposal to the CRTC for a system to distribute four Canadian TV services by satellite for reception and redistribution in the remote areas of the country, especially those north of the sixtieth parallel, using the underutilized facilities of the Anik A satellite. Hougen had been involved in local television production and distribution by cable on a small scale in Whitehorse since 1956 but had been hampered and frustrated by the lack of affordable program material in the far north. He realized that satellite distribution as it was then developing in the U.S. could provide an answer to this problem.

In 1979 the CRTC appointed a committee chaired by Real Therrien, one of the commissioners, to examine extension of television service to northern and remote communities. A special public meeting of this committee was held in Whitehorse in March 1980 at which Hougen, heading Canadian Satellite Communications (Cancom), discussed these proposals.

The Therrien Committee issued its report in July 1980 with recommendations which incorporated many of the suggestions made in the Cancom proposal, and in December 1980 the CRTC called for licence applications for a satellite service aimed at "remote and underserved communities." The Commission's definition of "underserved" covered those areas able to receive television service from not more than two broadcast stations, totalling an estimated 600,000 homes across Canada.

The first Canadian satellites, the Anik A series, A-1, A-2, and A-3, were launched between 1972 and 1975, and each had an operationl life expectancy of six years. Thus, early in this period it was necessary to consider the design of further satellites to replace these as they expired. In 1979 a second-generation satellite, Anik B, was launched. This had twelve transponders transmitting to earth in the 4 GHz band like the A series and intended to act as back-up for A-3, but in addition this satellite included four transponders transmitting in the 12 GHz band intended for use in experimental projects and were not to be used for any commercial service.

In September 1979 the Department of Communications, in conjunction with the equivalent Ontario ministry, began a joint experiment broadcasting direct to homes in remote areas of northern Ontario using one of the 12 GHz transponders on Anik B. This experiment was designed to test the technical and economic feasibility of using a third generation of domestic satellites, designed for downlink transmission in the 12 GHz band, for direct-to-home broadcasting.

A total of forty-six receive-only terminals were installed in various remote locations in northern Ontario where direct reception from regular broadcast transmitters was impractical. During the eight months of the experiment, they were able to receive the full program schedule of the TV Ontario network. The majority of these terminals, using 1.2 metre (four feet) diameter dishes, were installed at private homes. Three of them, using three metre (ten feet) diameter dishes, were installed at the headends of existing cable systems serving remote communities, and were used to receive the full program service of TV Ontario by satellite in place of the limited programming normally supplied by tapes circulated among the cable systems.

The reason why any form of direct-to-home satellite broadcasting service was not feasible using the Anik A satellites, although at that time they were underutilized, was technical. The very low power available at any receiving station from satellites operating in the 4 GHz band, coupled with the fact that this band was shared with terrestrial microwave systems and interference from these systems was difficult to avoid, made it essential to use antennas of at least twelve to fifteen feet in diameter together with very sensitive electronics. For this reason the installed cost of a TVRO earth station was typically between $20,000 and $30,000 and generally quite impractical for individual homes.

The 12 GHz satellites, because they used a spot-beam configuration, covering a smaller area, were able to deliver much higher signal levels at the earth stations, typically some twenty times more than that available from the 4 GHz satellites. Also the higher frequencies allowed the receiving antenna to deliver the same gain with an appreciably smaller diameter, considerably reducing the installed cost. In addition the 12 GHz band was not shared by any other service so interference was not a problem, making location of a station far easier. A further advantage of the higher frequency was the availability of more bandwidth so that these satellites could accommodate double the number of transponders, twenty-four instead of twelve on the earlier series.

The rapidly developing application of satellite technology to television distribution opened up possibilities, not only for access by individual households through direct-to-home broadcasting, but also for less expensive access by cable systems which were otherwise dependent on long-haul microwave systems for distant station reception.

Following the call by the CRTC in December 1980 for applications to provide a satellite distribution service to underserved areas, a public hearing was held early in 1981 and applications were received from several companies. Fourteen applications were considered but only four proposed the distribution of multi-channel television across Canada, and the major difference between these concerned the satellite technology to be used. In most licence application hearings the choice to be made by the Commission is between competing applicants taking into account ownership, financial ability, program proposals, and other competing factors. This hearing involved more than simply a choice between competing applicants, but rather a judgement between current and future technologies.

The dilemma arose because we were at the transition between two generations of technological development, a constant feature in communications arising from the rapidity of technical change. We are hardly accustomed to the use of one technology before another comes along which offers major advantages and we have to reorient ourselves and change our development plans to fit. This rapid development of satellite transmission was typical. We had only recently achieved an acceptable level of design and operation in using microwave systems for reception of distant TV stations, when along come satellites offering major advantages and able to transmit to many locations to which microwave was virtually inaccessible.

Although at that time successful experiments had been carried out with satellite distribution using 12 GHz from Anik B, the only satellites then in commercial operation were those using the 4 GHz band for downlink transmission, represented in Canada by the Anik A series. Thirteen of the applicants at the CRTC hearing proposed to use an Anik A satellite (later to be replaced by Anik D also using the 4 GHz band). Only one, North Star Home Theatre, proposed to establish what amounted to a DTH (direct-to-home) service using an Anik C series 12 GHz satellite, of which the first was due to be launched in 1982.

The Commission denied the application by North Star and awarded a network licence to a competing applicant, Canadian Satellite Communications (Cancom), and in so doing opted for the existing 4 GHz technology rather than the newer generation of 12 GHz satellites. It appeared that the CRTC was to some extent influenced by the fact that the first Anik C satellite was not due for launch for at least another year, and the Commission was anxious to extend service to underserved areas as soon as possible.

It was expected that, for cost-effective reception using the 4 GHz band, local distribution to small communities or clusters of homes from a shared earth station would be necessary either by cable or by low-power rebroadcast trnsmitters. Most of the applicants were reluctant to commit to cable for such small communities and therefore applied for the technical approval required to establish multi-channel low-power rebroadcast facilities. Cancom had been licenced to distribute programs from four Canadian broadcasting sources, and rebroadcast of all four from a centrally located TVRO required a group of four frequencies to be available, either in the VHF or UHF band. The VHF band was much preferred since equipment was more readily available and coverage of a given area easier to engineer. However, even in many relatively remote areas there was a lack of sufficient broadcast frequencies required for this purpose. As a result many of the applications could not be technically approved, and those that were acceptable would have been so delayed by the process that within a few months of approval the service could have been supplied direct to homes from the new Anik C satellite.

By early 1983, Cancom was clearly having problems with inadequate support of the service, while the Commission was expressing concern that, although 450 applications had been approved covering more than 850 communities, not more than seventy-five systems serving about 125 communities were in operation. In an effort to improve this situation, the CRTC in March 1983 approved the addition to the Cancom service of a package of four U.S. network signals; however, after these were added there were still only 268 communities receiving the Cancom service, a small percentage of the applications which had been approved.

By this time there were other unfavourable factors to add to the technical problems inhibiting licensees from proceeding with their plans to provide TV service to small and remote communities. Some had been discouraged by the time it was taking to process the applications — as much as eighteen months in some cases. Then a tight money situation made it increasingly difficult to finance these developments. In addition there were delays in the supply of the descrambling equipment required at each earth station.

During 1985 the government of Ontario conducted extensive research to determine the extent to which small communities were still underserved for television compared with larger communities. It found that approximately 180 communities totalling some 60,000 residents were receiving three or fewer television signals, and that a further thirty communities were receiving only the programs of TV Ontario, and these mostly by tapes circulated among them.

In 1987, in an attempt to improve this situation, the Ontario government initiated a capital expenditure program of $10.6 million to extend television service to these underserved areas. The program, known as TENO (Television Extension to Northern Ontario), provided one-time capital subsidies to private entrepreneurs and to existing cable systems to undertake system construction which would otherwise not be economically viable. In the majority of cases this construction involved new systems, including in each a headend designed for reception of the satellite signals distributed by Cancom, while in others it involved the extension of an existing cable system into adjacent underserved areas which had not been viable for inclusion in the original system design.

Some 240 applications were received and approved for capital assistance in cable system construction. In addition there were some thirty-five other applications for assistance in building a headend facility and distribution by low-power rebroadcast transmitters where the density of homes was too low to justify the cost of a cable system even with a subsidy.

Although the licencing of a satellite distribution service had not had the desired effect of improving television service to many of the underserved areas, it did increase interest in the possibilities offered by a direct-to-home service from the satellites using the new 12 GHz satellites. The CRTC was not anxious to make further moves in this direction at that time, particularly in view of the results of a study which had been commissioned by the Department of Communications in 1982. This study, entitled *Socio-Economic Impact of Direct Broadcast Satellites on the Canadian TV Broadcasting Industry*, had developed a statistical profile of the Canadian television industry, projected the possible impact of a DTH service on its various sectors, and concluded that the major impact might be a massive transfer of revenue from the CBC and its private affiliates to the private networks and independent stations.

In the meantime, in the United States the move in 1975 by Home Box Office of their pay-TV service to national distribution by satellite had precipitated a flood of similar services. This was encouraged by the Federal Communications Commission's open-skies policy on satellite ownership and operation, coupled with substantial deregulation of the cable TV industry in the U.S. Typical of the American free enterprise system there was no monopoly imposed on satellite ownership, and within a relatively short time several domestic satellites had been launched and were competing for business. Inevitably, when 12 GHz satellites became available, this business migrated to these with the result that, even though the first of them was the Canadian Anik C on which spare transponders were leased to the Americans, affordable direct-to-home reception rapidly became a reality only in the United States.

Since these satellites carrying American TV services could equally be "seen" from anywhere in Canada, it was inevitable that their direct reception in remote areas where little television was available would also proliferate. Some of these program services were pay-TV offerings similar to that of Home Box Office, but many were program services produced specifically for distribution by cable systems. However, they were all intended to be financed, one way or another, by the users at the receiving end, and most of them were scrambled or encrypted and could only be received by the addition of special equipment authorized from the transmitter. This led to the development of a so-called "grey market" in Canada, involving the sale of this equipment illegally imported from the United States or subscriptions to the services by the use of phony U.S. mailing addresses, eventually resulting in 1996 in legal action against some of the importers.

Early in 1993 the CRTC held a "Structural Policy Hearing on Broadcasting." It concluded that "DTH delivery of programming services via satellite will play a significant role in helping to achieve the objectives for the Canadian broadcasting system set out in the Broadcasting Act." To this end the Commission announced that it was encouraging interested parties to develop Canadian DTH services as a means of extending the delivery of broadcasting services to Canadians in underserved areas, to provide competition to cable undertakings, and to act as a made-in-Canada solution to the threat of unauthorized DTH services in the Canadian market.

In the fall of 1994 the government appointed a three-member panel to undertake a review of DTH policy. In April 1995 the panel issued a report recommending that "DTH satellite distribution undertakings" and "DTH pay-per-view programming undertakings" should be licensed by the CRTC. The Commission then issued a call for applications for these licences and received three applications for national satellite DTH distribution and six applications for DTH programming. These were considered at a public hearing in November 1995, as a result of which two licences were issued for national satellite DTH distribution, and five for DTH pay-per-view programming services. Of these five, two were for a national service in English, one was for a national service in

French, one was for an English service exclusively to western Canada (from Manitoba to British Columbia and the North-west Territories), and one was for an English service exclusively to eastern Canada (Ontario to Newfoundland).

In licensing these new services, the CRTC applied rules and obligations similar to those applied to the reception of programs for distribution by cable systems, including predominence of Canadian ownership, priority carriage of Canadian programming, simultaneous program deletion and substitution, and financial contributions to Canadian program production. The Commission did not propose rate regulation for these services since it considered that the degree of competition that the two new DTH distribution undertakings would face from each other, the cable industry, and other distribution technologies would create sufficient market pressure to discipline the rates charged to DTH subscribers. This was the first clear indication of the pending development of other licenced television distribution technologies competitive to cable.

In the early 1980s cable TV was still to most people what it had traditionally been for three decades since it first came into being: a medium for the improved reception of broadcast television signals, in fact a system of television reception and distribution which gave rise to and justified its original name, "community antenna television" or CATV. During this earlier period CATV systems were limited technically to the standard twelve channels in the VHF frequency spectrum used for TV broadcasting, since all receivers were designed to receive only these channels.

By the late 1960s pressure was beginning to develop for more channels on the cable, particularly in urban areas where there were a number of distant stations available which could add to program choice, while the presence of several local broadcast stations inhibited the use of their channel allocations on the cable. Technology soon catered to this demand with the introduction in the mid-1960s of transistors and solid-state equipment, and this enabled the development of systems with much greater bandwidth capable of handling frequencies outside the standard VHF broadcast spectrum. The introduction of a converter, which could be inserted in front of a subscriber's receiver to access these additional channels, released cable from the restrictions imposed by the TV receiver design and started a rapid progression in the number of channels which could be distributed by a cable system.

Parkinson's law clearly operates in the field of cable TV, just as it does in so many areas of human activity. If the channels are there they will be filled and a demand will be created for even more. So it was not long before the upper frequency limit of the cable equipment had been extended still further until by the early 1980s the potential channel capacity of a cable system had been increased from the original twelve to fifty or sixty. This increase in channel capacity enormously widened the scope of uses for cable TV to the extent that the distribution of television programs received off-air from broadcasting stations was becoming a smaller part of the total service package offered to subscribers. Many other video services could now be included in the overall service using both local and distant origination.

What was perhaps even more significant to the future of cable TV, particularly in a competitive environment, was the development of facilities on cable suitable for data transmission. This started, almost as a sideline, with pay-TV. This service, at least in some of its forms, requires an ability for subscribers to be able to communicate with the system head-end from which the programs originate, either to request programs, to be monitored and billed for them, or both. This in turn requires an ability by the cable system to transmit data in the upstream direction. There was also a growing interest by the industry at that time in the possibilities of providing a home security service to subscribers over the cable, and this requires data transmission between subscribers and the headend in both the upstream and the downstream directions. It was these two developing services which sparked an interest in data transmission over cable, but in considering these requirements it became apparent that a coaxial cable system has the capacity to accommodate an enormous amount of data, far more than would be needed for these specific services alone.

Data, in this context, comprises messages or information encoded and transmitted in the form of electrical pulses not unlike Morse code. This is known as the digital mode of transmission. Morse code was a very early and very crude form of digital transmission, originated manually instead of electronically, and hence transmitted at a very slow rate, typically some sixty words per minute. Data as we know it is quite capable of being transmitted at many thousands of words per minute, and a typical message could be sent in its entirety in a very small fraction of a second.

When an electronic medium such as cable is used for the transmission the speed depends on the bandwidth available. If a 6 MHz TV channel is used for this purpose this amount of bandwidth would allow a speed of something like a million words per second — a considerable improvement over the early systems of communication using Morse code. Security and some pay-TV services use a system configuration known as a polled circuit. This comprises a computer at the headend with a number of remote subscriber terminals and a common bi-directional communication channel between them. The computer controls the entire system, addressing each terminal in sequence and exchanging information on a time-shared basis. Since the address and enquiry message to each terminal is relatively short, and the information returned to the computer can also be short and precise, very little time is required for each polled address. Consequently, several thousand terminals can be addressed sequentially while occupying little bandwidth on the cable and still addressing an enquiry to each terminal every few seconds.

An extension of this concept can produce point-to-point circuits. As the term implies, in this case there is a necessity to transfer data, generally in both directions, between two specific points. In telephone technology a dedicated wire circuit is established between the two points, usually passing through one or more switching centres. Once the circuit is established it operates as a fixed circuit on a dedicated basis. The same configuration can be achieved on a cable system by the exclusive assignment of narrow frequency bands within the cable spectrum for each data circuit. The circuit passes through the headend, one frequency band being used upstream to the headend and a second downstream from the headend to the other terminal. A modem is provided to each subscriber to interface between the cable channel and the subscriber's equipment, and the necessary frequency changing is effected at the headend to interface the two channels comprising the complete circuit. As with the polled system, the bandwidth which can be made available on a cable system is such that a very large number of data circuits can be provided in each direction, thus making it possible to provide a service requiring two-way communication simultaneously with the distribution of one-way program services.

Just as text can be electronically transmitted in digital form so also can a video signal, although the amount of information, and therefore the bandwidth, needed for this purpose far exceeds the amount involved in a simple message.

In the early 1980s all broadcast signals were originated and transmitted in analog mode. The electrical output of a video camera is directly proportional to the intensity of the light falling on it from the scene being scanned, and this constitutes an analog signal, varying continuously in voltage level with the variations in the light. This signal can be scanned continuously and its level sampled at frequent intervals and defined by a series of digital numbers. When these numbers are transmitted in sequence and deciphered at the receiving end, the code they represent can be converted back to reproduce the voltage level of the original analog signal. The digital mode is not only advantageous for transmission; it can also greatly facilitate manipulation of the video signal for program production, for the insertion of commercials, and for other studio functions.

With the application of computer-based technology to broadcasting during the late 1980s it became more and more common for programmers to supply their signals to distributors in digital form, particularly after digital video compression (DVC) was developed. This technology drastically reduces the bandwidth required for a video signal to a fraction of that required for the equivalent analog signal and enables distributors to squeeze more video services into the space previously occupied by a single analog signal. Transmitting the broadcast services in digital mode also has another advantage. After a video signal has been digitized it consists of a train of pulses exactly like digitized data, separated only for recognition at the receiver by the carrier frequencies on which they are modulated. Subscribers' terminals can therefore be individually addressed by data accompanying the video, providing the facility for direct communication with the headend for program selection and control.

It was digitization and video compression which made feasible the establishment of direct-to-home broadcasting, enabling the successful applicants for these licences to include a menu of up to thirty television services, including several pay-per-view channels using a limited number of satellite transponders. Since these DTH services inevitably include a wide choice of popular American services in addition to the mandatory carriage of the Canadian services, they present a potentially formidible source of competition to established cable systems, even though they were intended when licensed specifically to provide television service in underserved areas.

To ensure a strong Canadian presence in the rapidly expanding communications environment, the CRTC has tried to licence as many affordable and widely demanded specialty services as distributors could carry. As of the late 1990s cable systems still rely primarily on analog technology which limits the number of services that can be distributed. With the deployment of digital video compression technology in cable TV the resulting increased channel capacity will enable systems to offer their subscribers a much greater number of services and, with the provision of addressability, will greatly facilitate the spread of services such as video-on-demand (VOD) which require two-way data communication associated with the video.

In 1990 a new television distribution technology emerged known as MDS or Multipoint Distribution System. This technology uses frequencies in a lower microwave band where sufficient bandwidth is available and, combined with digital compression, is able to accommodate upwards of sixty or more video services within the assigned 2.5 to 2.7 GHz band. These are then broadcast using a number of relatively low-power microwave transmitters to cover the area to be served. In this respect it is almost analagous to a cellular system used to provide telephone service within a limited area without the need for any wired connections to the subscribers. In fact the service has been referred to colloquially as wireless cable. A subscriber requires a small microwave antenna, similar to that used for satellite reception, and a wired connection from the antenna to the receiver, but no cable connection to the program source. MDS thus avoids the need for a cable distribution infrastructure such as that which forms the basis of cable TV.

In December 1991 the CRTC heard several applications for licences for MDS service in thirty-nine locations across Canada. It became evident at the public hearing that, while these applications focused on the use of MDS as a complement to cable, providing a television service in low-density areas where cable could not be viable, certain of the applications were predicated, at least in part, on operating in competition with existing cable. In view of this the Commission denied these applications and announced its intention to examine fully the impact of these changes and the future role of MDS.

Subsequently, the CRTC announced that it was prepared to consider applications for MDS licences as a means of providing a broad range of television services on a subscription basis to households in areas not served by cable. At the same time the Commission issued regulatory guidelines pertaining specifically to MDS undertakings, and indicated that the regulation of this new service, while taking into account the differences between the two technologies, would generally parallel that of cable.

Applicants were advised that MDS systems should not jeopardize the financial viability of neighbouring subscription-based systems. Moreover, the viability of MDS undertakings was not to depend on the provision of service to households located within the authorized service area of any licensed cable system, or within the principal market area of any other licensed system. The Commission was also prepared to licence joint MDS/cable operations in situations where MDS technology could more economically be used to extend service to households beyond the authorized cable service area, providing an applicant could demonstrate that such an extension would not unduly hamper or delay the provision of service by cable to these areas.

In October 1994 the government formally requested the CRTC to examine and report on the developing convergence of media and technologies, and the implications this would have for both Canadian culture and the national economy. This concern arose from the increasing public discussion of developments in the electronic movement of information — what was becoming known as the "information highway." The inquiry was to concentrate in particular on the nature of competition between cable companies and communications and other information providers, and the requirement that Canadian content have a prominent place on the information highway.

To this end the CRTC issued a notice of public hearing asking for submissions and comments from all interested parties. The Commission noted that the subjects to be addressed fell within the Commission's jurisdiction through two separate pieces of legislation, the Broadcasting Act and the Telecommunications Act. In response the CRTC received 1,085 written comments, and seventy-eight participants took part in a month-long hearing in March 1995.

In background information included with the notice of public hearing, the CRTC said that "the term 'Information

Highway' describes a network of networks that will link Canadian homes, businesses, and institutions to a wide range of services. The Information Highway will provide the necessary infrastructure for Canada's emerging knowledge-based economy, and therefore development of this highway will be critical to the competitiveness of all sectors of the economy. The building blocks of the Information Highway are the existing telecommunications, cable TV, broadcasting, satellite, and wireless industries. These industries are announcing plans to launch or develop new communications and information services, including content-based services. If these plans are implemented, and if there is sufficient demand, Canadians will rely more and more on these technologies for a wide range of services such as education, entertainment, social services, banking, and cultural products. Current discussions about these potential new services offer an incomplete picture of what the Information Highway will provide because many of the services that will be commonplace by the end of the century are not yet available, and many have yet to be conceived."

The Commission stated that one of Canada's strengths was that its cable TV and telephone industries both have networks that already reach nearly every Canadian. Together their local network facilities are capable of providing most of the advanced communications services that now exist, as well as many that will come to market in the next few years. Cable TV systems are broadcasting undertakings subject to regulation under the Broadcasting Act, and when they deliver non-programming services on the same facilities they use to provide broadcasting they also fall within the definition of a "Canadian carrier" under the 1993 Telecommunications Act.

In order to forward the objectives of the information highway it had now become the government's policy to encourage the interconnection and inter-operability of cable TV network facilities with those of other telecommunications service providers. Cooperation or sharing between cable licensees and telecommunications carriers would in future be permitted. The facilities and capacity of both cable and telecommunications carriers should be made available where practical for lease, resale, and sharing by service providers and other carriers on a non-discriminatory basis, and facilities, including support structures, should be provided in a manner that allows users to use and pay for only those parts of the network infrastructure that they require.

Clearly "convergence," as it was now being referred to, the melding of the various technologies which, in one form or another, provide the distribution of communications services, was recognized as a coming fact of life, encouraged and approved by government policy, which, before this sea change, had kept each confined and restricted.

The CRTC published the results of this inquiry in May 1995 in a detailed report entitled *Competition and Culture on Canada's Information Highway: Managing the Realities of Transition*. In this report the Commission said that new rules would be devised to remove barriers to competition resulting from the dominant position of telephone and cable companies, and to ensure that service providers have non-discriminatory access to various distribution systems by allowing more flexibility in how services are packaged and distributed. Fair and sustainable competition can best be attained by the creation of alternative networks for delivering services to subscribers. The Commission proposed to adopt measures that would ensure all telephone and cable subscribers have the freedom to connect the inside wire to the system of whichever distributor they choose, and it would be prepared to issue cable TV licences to telephone companies as soon as mechanisms were in place to give effect to competion in the local telephone business.

The dominant regulatory concern for the possibility of competition with other licensed systems changed following this study. Prior to this the CRTC's stated purpose in licencing alternative subscription-based distribution systems had been to avoid jeopardizing the financial viability of neighbouring systems. However, following this study, the Commission expressed the view that consumers should have increased choice among distributors of broadcasting and other services, and stated that it was prepared to consider licence applications by those (other than telephone companies) seeking to carry on distribution undertakings to compete in the core cable market. The Commission added that, consistent with this approach, it would cease attaching conditions to licences that have the effect of reducing competition between distribution undertakings and was prepared to accept applications for their deletion from all existing licences.

Thus ended the long period of protection by government from competition which had persisted, and indeed had been the foundation of Canadian government communications and broadcasting policy, since the earliest days of broadcasting. In December 1995 the CRTC issued a licence for the first Multipoint Distribution System (MDS) in Canada to serve various communities in Manitoba. This system comprised a total of nine low-power transmitters each carrying fifteen channels. Each of these channels was encrypted and used digital video compression (DVC), offering a total of forty video services, including pay-TV and pay-per-view, with the capability to expand to more than 150 as additional programming services become available.

Two of the transmitters were to be in locations where they would cover the cities of Winnipeg and Brandon respectively, both of which were already served by cable systems. This was consistent with the CRTC's recently announced policy of allowing, and even encouraging, competition between different distribution technologies. The announcement of the decision included the statement that "in the Commission's view the evolution of the Canadian broadcasting environment has been such that the role of MDS must now be seen as being competitive with rather than complementary to cable. In light of the dominance of the cable industry relative to other potential entrants there is no need to limit competition by other entrants in the broadcasting distribution market." Furthermore, the announcement added that "since DTH and MDS are both over-the-air systems which utilize addressable digital technologies they should have common regulatory requirements respecting the distribution of programming services." Apart from this element of competition with established cable systems, the applicant for this MDS licence estimated that there were 48,865 households within its proposed service area that did not have access to cable, so that the service would be serving its primary purpose of providing television to underserved areas.

One year later, in December 1996, a second MDS licence was issued, this time for communities in Saskatchewan. This system comprised a total of thirteen transmitters with all channels digitally encrypted, and offered an initial package of fifty video services capable of later expansion to between a hundred and 150. This system also included transmitters located in communities already served by established cable systems.

In August 1997 the CRTC issued a third licence for an extensive MDS system to serve virtually the whole of southern Ontario including the metropolitan Toronto area, but excluding the National Capital Region and surrounding areas in eastern Ontario. This system consisted of twenty-three transmitters offering a total of seventy-five video services with the capacity to upgrade channel capacity to 155 channels. As with the previous two MDS licences, this system also would serve substantial areas already served by cable, competing directly with these systems, as well as with DTH satellite systems licensed to provide service to surrounding rural areas not served by cable.

An aspect of this application which received considerable attention at the public hearing was the ownership of the successful applicant. Teleglobe, 24.3 percent owned by BCE Inc., held 74.5 percent of the voting rights in this operation. BCE-owned Bell Canada held a 58.5 percent interest in Telesat Canada and had the controlling interest in one of the licenced DTH satellite undertakings. In addition BCE was at that time conducting cable distribution trials in Repentigny, Quebec, and in London, Ontario, clearly preparing for the future entry of the telephone company into the cable TV business.

In their interventions the Canadian Cable Television Association, together with other cable interests, expressed concern that a telephone company might acquire control of an MDS undertaking soon after its licensing, giving the telephone company a major competitive advantage. In response to this concern the CRTC, in awarding this licence, made it a condition that all parties to the licence must apply to obtain the Commission's approval of any agreement or transaction that directly or indirectly results in a change in the effective control of the licence, or a shareholder in the licence, or in its controlling shareholder increasing its shares to more than 30 percent of the voting interests. The Commission added that in the event of such an application, it would examine the impact of the proposed changes upon the competitive environment, but in the absence of a compelling case to the contrary would not be disposed to approve such an application by BCE.

This particular approval was the first clear indication that the telephone companies, which for a long time had been eyeing the distribution of television by a parallel cable infrastructure, had at last got their noses in under the tent and would undoubtedly persist until they could participate fully in this service or even take it over, since they had never accepted any justification for two cable distribution networks.

In August 1997 the CRTC considered applications for MDS licences for areas in eastern Ontario and western Quebec, including the Montreal and Quebec City regions. Since the applicants included the group which had been successful in the last application, headed by Teleglobe, there was again strong opposition from cable interests based on BCE participation. This application was approved in February 1998.

In August 1996 the government had issued a statement of its policy framework on convergence, bringing to a close a series of initiatives aimed at introducing competition in virtually every area of the communications industry. The policy statement covered three major areas: facilities, content, and competition. It reinforced the policy objectives announced by the CRTC in 1994, clearing the way for cable and telephone companies to compete in each other's core businesses, while ensuring better access and strong Canadian content. It was intended to provide greater clarity and confidence for broadcasting and telcommunications companies as they entered each other's traditional areas of activity, and help consumers by providing more choices, with the assurance that Canadian content would remain prominent on their screens.

The policy was intended to allow fair and sustainable competition between cable and telephone companies. Telephone companies would be eligible to offer cable TV services once the rules for competition in local telephone services, including tariffs to enable cable companies and others to launch competitive services, have been implemented, and this may be done on a market-by-market basis. To ensure fair and sustainable competition, regulations would prevent cross subsidization and other forms of anti-competitive behaviour.

While new technologies allowed telecommunications and broadcasting companies to offer similar services, the distinction between the services would remain and would be guided by distinct regulatory systems. For example, when a telecommunications company provides broadcasting services those services will fall under the Broadcasting Act and its regulations; when a cable company provides telecommunications services they will fall under the Telecommunications Act and its regulations.

With due regard for differences in technologies, all broadcasting distribution undertakings would be expected to contribute to the production of Canadian programming and be subject to essentially the same rules governing this programming. However, if the undertaking itself wishes to provide broadcast programming the licence for this must be held by a structurally separate company, and in the case of a provincial crown corporation this licence will require an arm's length relationship and programming independence from governments.

In order to implement this policy the CRTC published in March 1997 a new regulatory framework for broadcasting distribution undertakings which set out the proposed rules guiding this competition. This regulatory framework was eventually incorporated into new Broadcasting Distribution Regulations which came into effect on January 1, 1998. The new regulations, which replaced the existing cable television regulations, apply to all distributors of broadcasting services in Canada, including cable, telephone, MDS, and DTH satellite services, and are intended to allow and foster fair competition between distributors and new distribution technologies. However, there were several aspects of this competition which, although mandated in principle, required additional consideration and discussion before their adoption in practice could be effected.

Probably the most important, and the most difficult to resolve, was the physical interaction between the distribution networks operated by the cable TV companies and those operated by the telephone companies because, unlike MDS and satellite DTH, both of these depend on wired entry from an external cable distribution system into the subscriber's premises. The telephone entry is by means of a pair of wires. Although this has been satisfactory for carriage of the audio-type services provided up to this point by the telephone companies, it is entirely inadequate for carriage of multi-channel video and similar services for which the wide bandwidth provided by coaxial cable is required.

The majority of households have two sets of inside wiring, a telephone pair and a TV coaxial cable. In many cases both are extended throughout the premises to serve extra outlets for each service, and they have generally been installed by the company providing the service. Few householders would be prepared to allow any of this to be duplicated in order to deliver the same service by a competing distributor. Clearly then the existing inside wire must be available to a competing service to make this service accessible to the user. Recognizing this, the CRTC included in the new regulations a requirement that, when a subscriber terminates cable TV service to contract with a competing distributor, the cable operator must offer to sell the inside wire to the subscriber so that he or she has the free and unrestricted use of it and can therefore connect it to an alternative service.

A cable TV operator proposing to add telephone facilities to its service offering does not necessarily have the same problem with entry into the home, since the technology used will permit the audio services to be delivered with the video over the coaxial cable. However, it does have another problem of a different nature. In order to be able to offer telephone service the cable TV operator has to have access to the switching equipment which enables the interconnection of subscribers, not only within their own neighbourhood, but to the world at large, and this equipment is located in the telephone company's central office and toll switching centres.

The Commission had already determined from earlier hearings that the development of effective competition in telecommunications markets requires the option of co-locating competitors' transmission equipment at the telephone companies' central offices to facilitate interconnection between them. It was difficult to determine a standard procedure for this since, although the dedicated use of floor space and access in each location would be ideal, it is far from practical in every case, and further discussions were required to arrive at a generally applicable solution and suitable tariffs to cover the costs involved.

The CRTC recognized that both these problems had to be satisfactorily resolved before effective competition could be in place to comply with the government's policy that no competing distributor should have a head start. Thus, competition between distributors, although legally effective as of January 1, 1998, under the new regulations governing all forms of distribution, could not in fact be in effect until some time after that date.

Another problem that arises from competition as it concerns the provision of telephone services by others is the matter of number portability. This is the ability of a telephone subscriber to retain his or her number in transferring from one service distributor to another. Clearly, if this were not available it would represent a considerable deterrent to making a change, however desirable such a change might otherwise be to the subscriber, and therefore not conducive to true competition. This was not covered in the new Broadcasting Distribution Regulations since it is a matter for regulation under the Telecommunications Act, but it is another of the problems which must be resolved before cable TV companies can be in a position to offer telephone services to their television subscribers, thus effecting true competition.

These network problems do not concern the competitive relationship between cable distribution systems and "wireless" systems — satellite DTH, MDS, and similar systems offering multi-channel video "over the air." In this case the outstanding problem which could affect the application of fair competition is the availability of sufficient channel capacity on the cable systems to carry all the services offered by the wireless competition. These systems take advantage of the much larger bandwidth available at the higher radio frequencies, and the greatly increased channel capacity made available very recently by the use of DVC (digital video compression). This technology will be applied progressively to cable systems to further increase their channel capacity, but this involves not only changes to the equipment in the distribution system, but more pertinently the addition of equipment at each subscriber's receiver to decompress the signals and convert them back to the analog form required by the receiver. Since this will require considerable capital expenditure it will be a limiting influence on the time required to make these systems truly competitive.

Taking all these problems into account, however, it appears probable that all aspects of the telecommunications industry will be effectively integrated by some time in the new millennium, and that there will be little distinction between the various services offered as far as the distribution medium is concerned. Furthermore, the competitive effects of this integration in the local distribution system to the subscribers will not be limited to the supply of television services. While wireless technologies are offering increasing competition to the established coaxial cable systems in the local distribution of multi-channel television, they are also beginning to represent a viable alternative in many situations to the POT (plain old telephone) service by paired cables which has been the prevalent, and almost unchallenged, technology for most of the last century. Deregulation of communications services, which has become an established fact throughout North America and other parts of the developed world in the last decade of this millennium, is resulting in growing competition with the established providers using a variety of technologies now becoming economically available.

The use of radio to deliver telephone service started with the development of cellular technology, which relieves subscribers from the necessity to be at the location of a fixed wireline terminal to initiate or receive a telephone call, and introduced the idea of communications mobility. This has proved to be a very popular service and has expanded rapidly throughout North America. However, regular telephone service to fixed locations continues to be provided by the wireline infrastructure, and so long as this service was the monopoly of regulated providers there was little incentive to adopt any replacement in established service areas.

The key developments which radically changed this situation in the late 1980s and early 1990s were the rapid proliferation of computer-to-computer communications, typified by the growing popularity of the Internet, together with the general deregulation of the communications services in North America and western Europe. Increasing computer communications demanded links of much higher speed and greater capacity to handle this traffic, while deregulation permitted access to subscribers by competing services. This introduced the necessity for access to subscribers without being dependent on the wireline connections owned and operated by competing providers, which in any case were inadequate for the new high speed requirements of the computer age.

The incentive for the development of wireless communications systems in the developed world, i.e. primarily North America and western Europe, is the need to be independent of the established wireline systems, with whom the new entrants are in competition, for entry into the subscribers' homes. In the developing world, where telephone penetration is far lower, the incentive is provided by the fact that wireless systems do not require the capital-intensive installation of a major cable network before service can be offered to the first subscriber, and the fact that wireless distribution is far more economic in sparsely populated areas where no hard-wired infrastructure exists.

It would appear, in the face of all these developments, that the early years of the new millennium, say by about 2020, will see a complete technical convergence of all forms of communications, be they voice, data, or video, into a single bit stream, indistinguishable from each other in the transmission medium, and each translated into its original form at the receiving terminal. It is this convergence of different services into a single data stream which is now referred to as "multi-media." The extent to which this convergence encompasses the ownership and operation of the individual services is a matter for politics and economics, and impossible to forecast at this time, but there is little doubt that the changes involved will be even more dramatic than those which have taken place in the last decade of the old millennium.

CHAPTER 9

Evolution of the technology

The development of the cable television industry has been followed as it evolved from the original basic service intended to overcome the problems experienced by viewers in receiving the signals from the nearest broadcast transmitter, to the point where it is providing a public service in delivering an ever increasing choice of programs from many sources. In fact it has become a valid alternative to the established broadcasting system.

Throughout this description there has been an effort to avoid technical terms or reference to the technical aspects so far as possible in deference to the average reader; however, cable TV is very much a technology-based industry. Through most of this history the various stages of this evolution have been technology-led; technical developments have made available services or improvements before there was an apparent need or demand for them, and the applications have then followed. This history would not therefore be complete without reviewing the evolution of the technology which has been the lifeblood of the industry's remarkable growth and development in little more than five decades, and this must inevitably be covered using the appropriate technical terms where applicable.

The very first reference to state of the art technology in this history was in the description in Chapter 1 of the earliest master antenna systems installed and operated by Rediffusion in multiple dwellings in London, England, in 1945. The ability provided by these systems to feed a number of television receivers from a single antenna was made possible by the use of an amplifier operating around the frequency of the off-air signal, 45 MHz, with sufficient bandwidth to accommodate the full 8 MHz (required by the British TV standards) of the transmitted signal, and sufficient gain to overcome the losses involved in distributing the signal through several hundred feet of coaxial cable and then splitting it to serve a number of outlets. This was no mean feat of design engineering for those days because 45 MHz was a very high frequency and 8 MHz an exceptionally wide bandwidth. Today, when bandwidths of several hundred megahertz operating up to at least 500 MHz are common, this would be considered a single-channel amplifier, but in 1945 it was very much state of the art.

The next technical development, also in the U.K., was the distribution of these frequencies on copper-pair cable designed and intended for transmission of audio frequencies, those not exceeding some 15 KHz, or 3,000 times lower than the lowest frequencies required for television transmission. As it turned out this was not a development which was in the mainstream of cable TV evolution because, although the technique was in use on a large scale for television relay in the U.K. for at least thirty years, it has now been abandoned in favour of coaxial technology with its much greater channel capacity. However, it does represent an important step in the evolution because a number of features which have subsequently become basic to the distribution of TV signals on cable were developed as part of this system.

The experimental system at the London Clinic, described in Chapter 1, was the first system in which the effects of "direct pick-up" were appreciated and measured. This is the type of interference which can give rise to a double image, or "leading (left-handed) ghost," on the TV screen when a local signal is distributed on the cable on the same channel on which it is received off-air. If the cable or the receivers are exposed to the off-air signal they can act like an antenna and, in the absence of special measures to avoid it, some of this signal can appear on the receiver screen with the cable-delivered signal. It will be displaced to the left of the main image because transmission delay through the cable will cause the wanted signal from the cable to arrive at the receiver later than the "ghost" signal resulting from the direct pick-up.

At the London Clinic it was possible to eliminate the interfering signal by installing a balancing transformer between the drop cable and the receiver input which could cancel out the unbalanced pick-up. The series of tests which resulted in this solution indicated the extent to which the TV cable, whether balanced or coaxial, is susceptible to interference from external signals but is shielded from them if affixed directly to a building. This susceptibility to interference from off-air TV signals is even more apparent in North America where the TV transmissions are horizontally polarized. Since most of the cable in the distribution system and in the drops is horizontal it is particularly prone to pick-up of signals radiating from transmitting antennas in the same plane.

In the TV distribution systems used at that time by London Rediffusion to provide an antenna service in multiple buildings, direct pick-up interference had not been a problem. These tests at the London Clinic showed for the first time that this was because all the cabling was attached to the buildings or run internally, and the adjacent structures provided a considerable degree of screening from the off-air signal. Later, when this technology was extended to larger relay systems and to coaxial systems erected in the open, direct pick-up became more of a problem requiring a change of channel for distribution over the cable if the ambient signal from the transmitter was strong. Even with the use today of much better cables it is still standard practice to convert the signals received from local transmitters onto other channels for cable distribution, and, if possible, avoid using the channels allocated to local broadcast transmitters for any video service.

Although the addition of television signals to radio relay cables already carrying audio signals had been proved to be practical, and the use of lower frequencies for transmission overcame the problems of high cable losses at broadcast frequencies and eliminated the problems with direct pick-up, further problems surfaced when attempting to use the existing relay cables in an outside environment. It was soon found that, even at the lower frequencies, the cable losses were subject to considerable variation, and it was proved by a series of tests that rain and dirt deposits on the cable, in addition to temperature changes, were the cause.

The effects of temperature were found to be pretty much as predicted from theoretical considerations, but changes in both attenuation and impedance due to varying environmental conditions resulting from rain and dirt deposits on the cable jacket were found to be even more significant. In particular there was a steady increase in the mean attenuation with time as the cable weathered. This was caused by radio frequency power loss in the surrounding dirt deposits, the magnitude of this loss being dependent on frequency and on the electrical characteristics of the dirt.

Following this discovery a copper or aluminum tape was added to a standard star-quad cable during manufacture, surrounding the pairs and beneath the outer jacket, to stabilize the characteristics against these environmental effects. Thus was born a new type of cable, a screened quad, or "squad" as it came to be called, and this became the standard cable for all new installations by the Rediffusion group intended to be used for relay systems combining both radio and TV services.

These environmental shortcomings of the standard radio relay cable meant that it was not in fact practical to add TV signals to the unscreened cables in existing installations; however, in most relay systems the introduction of television service in addition to the existing radio services required, or at least justified, upgrading the audio channel capacity of the system. This could best be done by installing a second cable, this time a squad, alongside the existing quad cable providing capacity for two television channels with their audio sound in addition to the two existing radio channels.

During the course of these tests coaxial cable of the type used in the London television distribution systems was used as a performance yardstick. This was a solid-dielectric cable with a copper-braid screen, very similar to the RG59 type cable in general use today on North American cable TV systems for subscribers' drops and inside wiring. It was found that this too experienced increases in attenuation with weathering, although the increases were not nearly as great as those on the radio relay cable. This increase was due to the screening effect of the braid being far from perfect, resulting in some electrical interaction between the inside and the outside of the screen. Part of the increase may have been due to a reduction in conductivity of the screen caused by oxidization of the braid wires with aging.

It was this early experience which resulted in the virtually standard use of solid-aluminum sheathed coaxial cable throughout the cable TV industry in North America, with the added precaution of converting the signals from local broadcast stations to other channels for distribution on the cable to avoid "leading ghost" problems due to residual direct pick-up interference. The use of this type of cable also makes a significant contribution to reducing the extent to which any radiation from the cable interferes with essential radio services, a form of interference which is the subject of strict regulatory limits.

The first large-scale cable system designed specifically for television distribution using coaxial cable was that by Rediffusion in Montreal (described in Chapter 2). With this one exception, the first coaxial systems built in the United States and Canada were to all intents and purposes extensions of the master antenna principle already used in the apartment buildings in London. They were generally located in smaller communities which, by virtue of their size and distance from the nearest major city, had access to only one broadcast station. Indeed, it would be this relative remoteness, coupled with a market demand for access to the new television service, which was the reason for such an entrepreneurial venture. The signal would be received on a special antenna, then amplified and distributed on the same channel over cable equipped with single-channel amplifiers.

Where reception from more than one broadcast transmitter was available, such as in the apartment installation by Jerrold Electronics in New Jersey (referred to in Chapter 3), single channel strips were linked together to handle up to three non-adjacent channels. The earliest of them used amplifiers which had been developed as antenna amplifiers for use in multiple-dwelling MATV systems. They soon demonstrated their inadequacies for distribution over more than a few hundred feet of cable, because they had not been designed with cascading of several amplifiers in mind, and noise and distortion products built up rapidly to intolerable levels after only a few in cascade. It soon became evident that the design of these amplifiers had to be substantially improved, and it was not long before amplifiers were in production which were more suitable for these CATV applications.

Fortuitously, the same design elements which were essential for these improvements in single channel performance also made it possible to widen the bandwidth. It was not long before amplifiers were available which were capable of carrying the entire low VHF band — channels 2 through 6 in North America — requiring the reasonably flat transmission of frequencies from 54 to 88 MHz. This coincided with the availability to some of the early CATV systems of two or more broadcasting stations, and the need to accommodate more than one channel on the cable systems with distribution over longer distances to serve more subscribers in larger areas.

Although at this point these systems could carry all five channels in the low band, television receivers at that time had not been designed with tuners capable of separating adjacent channels because this was a condition intentionally avoided in the original allocation of broadcast frequencies. To cater to this limitation it was necessary for the cable systems to limit their carriage to three non-adjacent channels of the five, generally 2, 4, and 6, but at least they now had a three-channel capability instead of only one.

Although the ability to carry three channels was a major advance, it was not sufficient for some cable systems operating in areas where three broadcast stations were receivable off-air with domestic antennas while more distant stations could be received and distributed by cable. Cable TV has always been driven by the necessity to be able to deliver not only better and more reliable reception than could be obtained using domestic antennas, but more distant stations if these were receivable using the superior technology inherent in a cable TV headend.

So long as the adjacent channel limitation of the receivers existed this could only be accomplished by extending the bandwidth of the amplifiers beyond 88 MHz, the upper limit of channel 6. Immediately above this frequency there are several bands in the North American system allocated to FM radio, communications, and other services, and the next TV channel, number 7, starts at 174 MHz. The TV allocations then proceed up from there to channel 13, the upper edge of which is at 216 MHz. This group of seven channels is known as the high VHF band. There is a further group of TV channels extending from 470 MHz to 890 MHz, comprising channels 14 to 83, which are referred to as the UHF band, but at this stage in the evolution these frequencies were much too high to be considered as useable for distribution on cable, since the practical upper frequency was limited by the available amplifier design.

The next development in amplifier design therefore had to be the extension of the amplifier bandwidth from 34 MHz (extending from 54 to 88 MHz) to 162 MHz encompassing the entire VHF band from 54 to 216 MHz. This increased the capability of a system from three channels to seven, allowing for the inability to use adjacent channels, or to twelve channels if this limitation in the receivers could be overcome.

Amplifier development reached this stage by the late 1950s and was undoubtedly a major reason for the considerable growth in the number of CATV systems under construction by this time. This was particularly the case in the U.S., aided by the steadily increasing number of broadcasting stations, many of which were not affiliates of the three networks and were therefore offering alternative programs in competition with the network-affiliated stations. The competing stations were generally distant stations to viewers in the larger communities, and cable was the only way their signals could be adequately received, but they could only be accommodated on cable if the capacity could be expanded beyond the three channels in the low band already used for distribution of the network stations.

The problem with adjacent channel discrimination on the receivers centred on the sound carrier of the lower adjacent channel to the one being received. The television standards used in North America were first adopted by the FCC in 1941, based on recommendations of a National Television Systems Committee (NTSC); for this reason they are known as the NTSC standards. In these standards the bandwidth of the video signal produced at the source is 4.5 MHz. When this is modulated onto a carrier for transmission it produces two sidebands of this width covering a band 9 MHz wide centred on the carrier. Since all the intelligence in the signal is contained in each sideband it is only necessary to transmit one of the two, and this will easily fit into a broadcast channel 6 MHz wide.

There are complications involved in single-sideband transmission, and to avoid costly additions to TV receivers to accommodate these a technique known as vestigial sideband transmission was adopted. The output of the modulator is filtered, so that only one sideband is fully transmitted while the other is restricted to 0.75 MHz on the other side of the carrier with a special sloped response which simplifies filter design in the receivers. In this way the video signal is transmitted within a total bandwidth of 5.25 MHz leaving 0.75 MHz of the 6 MHz channel to accommodate the sound carrier and adjacent channel spacing.

The sound carrier on the lower adjacent channel is located 1.5 MHz below the video carrier of the wanted channel but, although it was standard practice to broadcast it at a level 6 decibels (db) below that of the video carrier, the vestigial sideband filters in the receivers were not able to provide sufficient rejection of this carrier. It was soon discovered that on cable, if the sound carrier were to be reduced in level by a further 6db this interference with the upper adjacent channel could be eliminated allowing the use of adjacent channels on standard receivers. This could be done fairly easily at a cable system headend, thus further increasing the capacity of a system designed to carry the two VHF bands from seven to the full twelve channels, and this is now a standard feature on all cable systems.

It appeared that the ability to distribute up to twelve channels on a cable system was likely to meet any demand in the foreseeable future, particularly as few cable systems in the U.S. had ready access to more than three broadcast stations affiliated with the three major networks, and perhaps two or three other stations which could be considered as alternative sources. However, by the mid-1960s the number of stations which were independent of the networks was increasing, and every effort was made to receive one or more of these for distribution on the additional channels now available.

Since these were frequently distant stations this led inevitably to progressively more sophisticated developments in antenna design for use at CATV headends. The antenna commonly used from the earliest days for distant station reception had been the yagi, an antenna with good gain and directivity, which could be improved further by using several stacked either vertically or horizontally in an array. An improvement on the yagi antenna was the log-periodic, which was basically similar but had a broader bandwidth and was useful where more than one channel within the same band had to be received from the same direction.

The parabolic was another type of antenna developed at this time for use in cable systems which were really reaching for more program choices from distant stations. This form of antenna is commonly used nowadays for satellite and microwave link transmission and reception; however, these systems use frequencies above 2 GHz, at least ten times higher than those used for VHF television broadcasting. Since the size of an antenna is inversely proportional to the frequency at which it is operated a parabolic antenna at TV frequencies, especially on the low VHF band, has to have physical dimensions twenty to forty times larger than those of a satellite or microwave antenna operating in the 2 GHz microwave or satellite downlink bands.

Such an antenna, designed to operate in the VHF television band, comprised a set of some eight to ten towers about sixty feet high, each shaped vertically to a parabolic curve and spaced about thirty feet apart along a parabolic curve on the ground, with a screen of closely-spaced wires suspended between them. This screen formed a parabolic reflector which received the distant signals and concentrated them onto a yagi antenna array located at the focal point of the parabolic curve some 150 feet in front of the screen.

The parabolic antenna had a high gain of the order of 30db in the low band and 40db in the high band and very good directivity, and was capable of bringing in television signals from transmitters a hundred miles or more away, well beyond line-of-sight. This could be achieved by using the tropospheric-scatter mode of propagation, in which part of the transmitted signal radiated up towards the troposphere is reflected down to the receiving antenna. This type of antenna, however, required several acres of ground to accommodate the towers with their supporting guys, together with the focal-point structure. Consequently the number of locations where it could be built, not to mention the considerable cost involved, limited its use to a relatively small number of systems in which this was the only available means of bringing in the distant stations needed for a viable cable system in the community.

One characteristic of tropospheric propagation which tended to limit the use of the parabolic antenna was the presence under certain climatic conditions of very rapid fading. The incidence of deep fades was different at receiving locations only a few hundred feet apart across the direction of the signal, so that as the signal faded on an antenna at one location it might still be at maximum strength on another not far away. This phenomenon, known as space diversity, could be used advantageously by erecting two similar antennas spaced 50 to 100 wavelengths apart at the frequency of the signal being received (1,000 to 2,000 feet in the low VHF band), with fast solid-state switching to select the best signal at any moment. This arrangement was capable of substantially improving the reliability of the received signal and raising it from commercially unuseable to acceptable. However, such an antenna system was relatively expensive to install, required a considerable amount of space, and could only be justified in a few cases where inclusion of the station concerned was necessary to ensure the viability of a cable system in a large community which would otherwise be incapable of supporting a system.

Although large parabolic antennas and other types of sophisticated antenna arrays mounted on tall towers provided access for many cable systems to otherwise inaccessible distant stations, they gave way in the late 1960s to a more reliable, even though more costly, means of access — long-haul microwave. In Canada up until 1968 under the licensing agency of that time, the Department of Transport, there was a rule that the receiving antennas for a cable system had to be within ten miles of the area being served. Only in exceptional circumstances were remote receiving sites licensed with a microwave connection to the service area, and these were invariably for access to Canadian stations by communities not otherwise served by the Canadian broadcasting system.

In 1968 the regulation of cable TV in Canada was included with broadcasting under the mandate of the newly created Canadian Radio-Television Commission (CRTC) as described in Chapter 7. In 1970 the Commission developed new policies on the importation of distant broadcast stations, including the use of microwave for the remote reception of U.S. stations. As a result the use of sophisticated antenna systems for long distance reception declined in favour of microwave where the costs involved in this technology could be justified, since the signal quality and reliability were greatly improved.

Long-haul microwave uses facilities provided by telephone companies, either on existing toll routes or by dedicated construction between specific locations. This type of delivery is costly to provide and maintain, and generally could only be justified if it could be shared between a number of participating cable systems, each with sufficient subscriber potential to support the costs involved. Nevertheless, there were several such systems in Canada, including one serving cable systems in four Atlantic provinces which was some 1,200 miles in length from the headend near the New Brunswick-U.S. border to the furthest participating system in the southeast corner of Newfoundland.

This was not the only application of microwave technology to be applied to cable TV. Early in the 1960s a multichannel microwave system termed AML (amplitude modulated link) was developed in the United States which was capable of transmitting a full complement of TV channels on a single carrier in the band 12.7 to 12.95 GHz. This band, allocated by the Federal Communications Commission for the dedicated use by CATV, became known as the Community Antenna Relay Service (CARS) band. These frequencies are too high for long-haul use requiring multiple repeaters due to high propagation losses and the susceptibility to severe fading in rain. They are, however, well suited to use on single-hop systems not exceeding fifteen miles in length, and can be applied to many situations where such a link can replace ten to fifteen miles of cable from a remote headend into the distribution system in a community. Furthermore, since the available bandwidth of 250 MHz is sufficient to accommodate some forty 6 MHz channels, or the full VHF band from 54 to 216 MHz as carried on cable, it was ideal for a broadband microwave system such as the AML development.

AML stands for amplitude modulated link, so named because the transmitter accepts the full band of frequencies as carried on cable and amplitude-modulates it onto a single carrier in the 12.7 to 12.95 GHz band. At the microwave receiver this signal is demodulated back to VHF and the full range of channels is available at the output for onward transmission by cable exactly as they were presented to the transmitter, so that the link appears in effect to be an extension of the cable itself. The use of AML increased rapidly in the early 1980s, particularly in the United States where it was included in the proposals for new franchises in many major cities as the primary feed in a city-wide distribution system using a central headend and cable distribution from several microwave-connected hubs.

The developments in cable technology described earlier in this chapter were concerned primarily with extending the bandwidth of the line amplifiers to accommodate more channels. The coaxial cable itself is a broadband device with no inherent bandwidth limitation except for the attenuation which increases with frequency, so that the limitation is imposed primarily by the amplifiers which must be included about every two thousand feet in the trunk to overcome these losses, and by the bandwidth of the passive networks which must be included in the cable for the purpose of feeding signal to subscribers or to other feeders.

Once amplifier bandwidth had been extended to 216 MHz, the upper edge of the VHF band, and the adjacent channel problem in the receivers had been overcome by reducing the level of the sound carriers, cable systems were capable of delivering up to twelve channels on the assigned VHF broadcasting frequencies. While this provided 162 MHz of bandwidth between 54 and 216 MHz, the limits of the assigned VHF band, only 72 MHz of this was occupied by the twelve broadcast channels, and in looking for additional channel capacity some of the unused portion of this bandwidth was too attractive to be ignored.

The gap in the original broadcast frequency allocations between the top of channel 6 at 88 MHz and the bottom of channel 7 at 174 MHz was intentionally introduced in order to avoid second harmonics of the low-band carriers. These harmonics, at twice the frequency of the originating carrier, could be generated as spurious signals in a low-band broadcast transmitter and would appear between 110.5 MHz and 166.5 MHz, where they could interfere with any TV carriers present in that band. Harmonics are generated in an amplifier if the circuitry is non-linear and these can be reduced by the introduction of "push-pull" circuitry.

The push-pull type of amplifier circuit has the ability to amplify the input signals while cancelling any second harmonics generated within the amplifier. Thus, if additional signals on frequencies between 88 and 174 MHz are included at the input of an amplifier designed for the band 54 to 216 MHz they will appear at the output free of any interference from second harmonics which may be generated within the amplifier by the low-band channels. This circuit technique was first introduced in 1957 in amplifiers designed to cover the band 7 to 95 MHz at a time when the use of frequencies below 54 MHz were being considered to take advantage of the lower cable losses at these frequencies.

In using the band between 88 and 174 MHz it was expedient not to use that part between 88 and 126 MHz because this allocation accommodates the FM broadcast band and other radio services of an essential nature which could be subject to interference. By using the portion between 126 and 174 MHz room was created on the cable for a further eight channels, thus making feasible a twenty-channel capability. This stage in cable evolution was reached with the introduction of the first commercially available push-pull amplifiers about 1968. The design changes involved in this development were such that it was not too difficult to extend the overall bandwidth of the amplifiers by a further 80 or 90 MHz, providing performance up to 300 MHz and capacity of the order of thirty channels, and this was achieved by 1973. Indeed this is the nature of technical evolution — each progressive development makes the next step that much easier, and this has certainly applied in the design of cable TV line amplifiers to the point where this equipment is now being developed with bandwidths approaching 1,000 MHz.

While the bandwidth of cable TV amplifiers was being steadily expanded, the design of the associated passive equipment had to keep pace. A high frequency transmission path must maintain a matched impedance condition at all times to avoid destructive reflections and loss of signal power which will occur wherever there is a mismatch. Cable TV systems use coaxial cable which has a characteristic impedance of 75 ohms, and it is necessary that every piece of equipment inserted in the signal path also have the same input and output impedances within very close limits at the point of connection with the cable.

If a cable is to be split into two directions this impedance match condition cannot be maintained if the two cables are literally connected in parallel. So they must be connected through a passive network designed to provide two outputs while presenting an impedance of 75 ohms to the cable at all ports. This impedance matching must also be achieved at every point at which the cable is tapped to feed a subscriber without taking too much from the ongoing feeder. In the early days these passive networks were designed using resistors, but these were soon replaced by transformers which were capable of meeting the impedance requirements with much less internal loss. However, unlike a resistor, a transformer has reactive components and is therefore sensitive to frequency. Thus, as the system bandwidth was extended to progressively higher frequencies it was necessary to modify the design of these passive components so that their frequency responses were reasonably flat up to the same limits while maintaining the same impedance match.

Coincident with these developments in equipment design there were also considerable improvements in the design of the coaxial cables used for these systems. The early systems were designed and built around the RG series of coaxial cables which had been designed for military use during World War Two. These consisted of a copper centre conductor surrounded by a dielectric material of solid polyethylene, with an outer shield of braided copper wires in turn covered by a plastic protective jacket. This comprises a reasonably efficient coaxial arrangement, although the porous nature of the outer shield makes it susceptible to leakage of radio frequency energy. This could allow radiation of signal from high-level portions of the system, and conversely interference with the system from signals present outside the cable. This unwanted transfer of energy was minimized by applying two separate braided shields — the so-called double-shielded arrangement.

Around 1961 a foamed dielectric was introduced in which the polyethylene dielectric was impregnated with a mass of tiny air bubbles giving it the consistency of a sponge. Since dry air is the most efficient dielectric this had the effect of reducing cable loss still further; however, the simultaneous introduction of these two features did not attain the maximum theoretical improvement because the foamed dielectric tends to act like a sponge if exposed to moisture, and this in turn increases the cable loss again. In order to avoid this it is necessary that the outer shield should be homogeneous so that it acts as a water vapour barrier and prevents the entry of moisture.

Attempts to modify the tape-screened cable by soldering the overlapping seams were not entirely successful. Finally around 1964 coaxial cable using an outer sheath consisting of a solid-drawn aluminum tube, which had originally been designed for the Rediffusion system in Montreal, provided the answer. This type of cable quickly came into general use for CATV construction, and most modern systems, at least in North America, are designed on the basis of solid aluminum cable together with solid-state equipment.

Expanding the capacity of the cable system to carry more than the twelve VHF broadcast channels shifted channel limitation again to the receiver, since at this time most receivers were limited to reception of the allocated broadcasting channels. This was overcome by the introduction of the subscriber's converter, a device external to the receiver, to be installed in the drop cable at the TV set input. This device has the ability to receive all the channels transmitted on the system, even when in excess of the standard allocations, and converting the channel to which it is tuned to either channel 3 or 4 in the low VHF band, since one or other of these will be free of any local broadcast allocation wherever the set is located, the set then being left tuned to this channel. This transfers the channel tuning function from the set to the converter and so removes the limitation in the subscriber's equipment and puts the ball back in the distribution system's court for further extension of channel capacity. In fact, with the introduction of this device, the limiting factor once again became the amplifiers. Later the converter was incorporated into the TV set by many manufacturers, providing what came to be known as a "cable-ready set" capable of tuning to any channel on the cable system without the need for an external converter.

A basic characteristic of broadband amplifiers carrying several channels is the generation within the amplifier of "noise" and distortion. The former, if excessive, causes a picture on the receiver to appear snowy, while the latter gives rise to various visible imperfections in the picture depending on the type and magnitude of the distortion. Noise is visible at the end of the system if the signal level at the input to each amplifier is too low in relation to the noise generated within the amplifier. Distortion is visible if the signal level at the amplifier outputs is too high, causing excessive generation of distortion products due to non-linearity in the amplifier gain characteristics. Thus, acceptable levels of noise and distortion depend on a careful balance of amplifier input and output levels. When these amplifiers are operated in cascade, as they must be to extend the reach of a TV cable, these effects are cumulative, and this imposes a limit on the number in cascade and therefore on the length of the cable. Furthermore, the distortion generated in each amplifier is proportional to the number of channels being carried, and for this to be within acceptable limits at any subscriber's receiver the number of amplifiers operating in cascade must be limited.

The continent-wide distribution of television program services to cable systems by satellite began in 1975 with Home Box Office pay-TV. By 1980 the number of these services was rapidly increasing. This effectively removed any "distant station" limitation from cable systems and made many more services available for distribution. Consequently, the demand for more channel capacity escalated. The design of cable TV amplifiers was then concentrated, not only on the extension of broadband technology to accommodate more channels, but at the same time on improvement of the amplifier designs to reduce the distortion generated by the increased channel loading and thus permit this extra capacity to be used without degrading the overall performance of the system.

A major development in the evolution of the cable TV line amplifier occurred around 1963 with the introduction of the first commercial amplifiers using transistors in place of vacuum tubes. From the beginning of cable technology these amplifiers had incorporated tubes, and all the associated circuit components such as capacitors, resistors, etc., were discrete components, just like the television and radio sets of the time, and indeed all other electronic equipment. As a result the amplifiers, which had to be installed outdoors, generally on the poles and exposed to the elements, were mounted in large sheet steel cabinets for adequate protection and these were both bulky and heavy. Furthermore, since the tubes required a substantial amount of power, each amplifier had to be connected to a source of metered primary power at the location of the installation.

With the introduction of transistors, and the associated development of other solid-state devices, it became possible to miniaturize an amplifier to the point where all the components of the circuit could be included in, or mounted on, a printed circuit board, eliminating the need for inter-component wiring. This drastically reduced the size of an amplifier and, together with the basic nature of a solid-state device, produced several revolutionary results.

First, the miniaturization and the virtual elimination of internal wiring drastically reduced stray capacitances and inductances which were the main limitations on bandwidth expansion at these frequencies, and thus permitted the evolution of real broadband design. Second, the much lower power requirements of solid-state components, and the fact that they do not require the high voltages necessary to the operation of vacuum tubes, led to the development of power feeding over the coaxial cable. This in turn permitted the design of compact amplifier housings which could be hung on the strand and, to all intents and purposes, were incorporated as part of the cable. Third, these two features further broadened the limits of circuit design which permitted more complex circuitry to improve the distortion characteristics and so increase amplifier capacity.

The impact of solid-state electronics in cable TV line amplifiers was sufficient to revolutionize the design and planning of cable systems. The first of these amplifiers became available for commercial use in 1958. By 1965 all equipment suppliers had a full line of amplifiers using this technology, and all new systems were being designed and built, and many existing systems rebuilt, using this equipment. By 1970 tube-type equipment was virtually obsolete and no longer manufactured since solid-state equivalents were available as direct replacements. These features all revolutionized cable TV amplifier design to the extent that by 1971 thirty-five channel capability had been achieved with the upper frequency limit extended to 300 MHz.

During the early years of CATV development in the 1950s and 1960s government control of the industry was minimal, and consisted largely of defining certain technical standards to be observed in operating the systems. These were spelled out in Radio Standards Specification No. 102 issued in March 1956 by the licencing agency at that time, the federal Department of Transport, as described in Chapter 7. The technology, such as it was at that time, was so new that these were hardly standards in the generally accepted sense of the word. By 1968, when the new Broadcasting Act designated cable as part of the broadcasting system and licensed as such by the newly established CRTC, the need for more specific technical controls was becoming very evident.

The purpose of any control over the technical operation of a cable TV system is to ensure that, as one link in the broadcasting system, the cable will not unduly degrade the system as a whole. This requirement is a normal one in any communication system comprising a number of links in tandem. It is necessary to specify a certain standard of performance for each link in order to attain some other standard, necessarily lower, for the system as a whole.

A typical example of this is in the telephone system. Certain minimum standards must be set for overall subscriber-to-subscriber loss, signal-to-noise ratio, and other parameters to achieve a connection over which intelligibility and quality of speech are acceptable. However, if the local central office network were designed to this standard then a connection from a subscriber in one city to a subscriber in another over the long distance network would probably not be acceptable, and international calls involving a third network might be quite impossible. The total allowable degradation must therefore be apportioned between the various elements of the system which could be connected in tandem, and each must inevitably be better than that required of the whole.

Applying this analogy to cable television we find at least three basic elements present in all systems: the broadcaster's equipment, including the transmitter and everything behind it — video processing equipment, studio-to-transmitter links, and in many cases a long-distance network; the cable distribution system from headend to last subscriber; and the subscriber's receiver. In many cases there is a fourth element, a distribution system between the transmitter and the headend, consisting of microwave or satellite links.

The original "standards," such as they were, for the broadcasting system as a whole evolved with a reasonable apportionment of allowable degradation between the broadcaster and the home receiver with no allowance for any intervening system. It has always been easier to enforce these standards on the broadcaster since this part of the system is subject to licencing and, through it, control of the technical performance. By 1968, when cable TV in Canada was brought under the licencing authority of the CRTC, it was evident that Radio Standards Specification No. 102 was quite inadequate for this purpose and the newly formed Department of Communications (DOC), which had taken over the technical supervision of the industry from the Department of Transport, started to develop a more complete set of performance standards.

In April 1969 DOC issued a draft of a new standard, "Broadcast Procedure 23: Minimum Performance Standards" (BP23), for industry study and comment. This did indeed receive exhaustive study by the technical committee of the Canadian Cable Television Association, the Electronic Industries Association representing manufacturers of TV sets and cable equipment, and the television committee of the Canadian Radio Technical Planning Board comprising representatives from the cable TV industry, receiver manufacturers, broadcasters, common carriers, and broadcast consultants. Following considerable amendment as a result of these studies DOC finally issued Broadcast Procedure 23 on March 29, 1971, to be effective for all new systems from July 1, 1971, and to apply to all existing systems progressively over the following five years.

Subsequently, on August 6, 1971, DOC issued a supplemental document "Broadcast Procedure 24: Proof of Performance Procedure for Cable Television Systems" (BP24). This document described acceptable methods for evaluating the performance of cable TV systems and outlined details of the technical submissions required. These included proof of performance before any new system could be licensed for operation, and, over a period of five years, a proof of performance on every system already licensed and in operation.

The impact of these new technical standards was considerable, particularly on existing systems. For new systems the expense involved in preparing a technical brief and undertaking the subsequent proofs of performance as part of the licence application procedure was nominal in relation to the costs involved in constructing the system, but for existing systems it was a different matter. Many of the worst cases were older systems where the owners may well have been considering acquiring more modern solid-state equipment if only to reduce maintenance costs. There was little doubt that the standard of service provided to subscribers on some systems left much to be desired, either because of inadequate original design, outdated equipment, or inadequate maintenance. Completion of the BP23 and BP24 program had the effect of upgrading cable systems as a whole, permitting them more readily to add channels or other services, and went a long way to effecting the conversion of the industry from CATV to the more sophisticated concept of Cable TV.

The trend to increasing bandwidth has continued. By 1996 the upper frequency limit of commercially available amplifiers had been extended to 860 MHz, capable of handling 120 TV channels each 6 MHz wide, while still meeting stringent performance standards with twenty or more operating in cascade. The demand for more and more channels is insatiable, and is now fuelled by more than a desire for additional program choice as it was consistently during the first three or four decades of cable's evolution. Other demands are now being made on the channel capacity of cable systems related to the needs of other types of service in addition to the distribution of broadcast television. Quality and reliability of the service are becoming increasingly more important, or perhaps more to the point, cable operators now recognize that these features are of as much concern to the subscribers as the number of channels or the amount of program choice.

Quality is a function of a number of factors inherent in the distribution of television signals by cable. It is determined first of course by the quality of the originated picture, and the reproduction on a subscriber's receiver after transmission through the medium, whether broadcast or cable, cannot be any better than this. Transmission through a cable system inevitably results in further deterioration of picture quality, mainly due to the noise and distortion introduced by the amplifiers through which the signal must pass, plus accumulations of small reflections in the system caused by slight impedance mismatches at the many junctions between cable and equipment. Thus, the overall quality is improved if the noise and distortion levels in the individual amplifiers are reduced, and is further improved if the number of amplifiers in cascade can be reduced. Furthermore, this will contribute to a reduction in the number of cable to equipment junctions, and so further improve the quality by minimizing the number of cumulative reflections occurring at these points.

Reliability is a measure of the continuity of the service, in terms of frequency of failure due to whatever cause, and in the event of an interruption in service, the length of time before it can be restored. Failure of service may be due to a discontinuity in the cable or failure in any of the electronic equipment included in the signal path. Short of an actual break in the cable itself a discontinuity is generally located in one of the many connectors which form the junctions between the cable and the equipment. A failure in electronic equipment generally implies a component failure in an amplifier or loss of primary power to an amplifier. Clearly, reducing the number of amplifiers in cascade will reduce the potential for these types of faults and improve the reliability of the system as a whole.

Another way in which the number of amplifiers affects reliability is related to the requirements for primary power supply. The power required by a modern solid-state amplifier is minimal — typically some 20 to 60 watts, depending on the type of amplifier — and this power can be fed over the coaxial cable at voltages which for safety reasons do not exceed 60 volts, although extension of this limit to 90 volts has recently been approved. The power unit, which takes primary power at 110 volts A.C. and converts it to 60 volts for feed over the cable, will have a capacity of at least several hundred watts and is therefore capable of providing power to a number of amplifiers, typically some ten to twenty. Battery standby facilities can be provided at the location of a power supply unit which, in the event of a failure of the primary power, will ensure continuity of service at least for sufficient time to enable restoration of the regular supply or installation of more long-term standby facilities. Clearly, the fewer amplifiers there are in the system the less is the chance of service failure due to loss of power. At the same time there will be fewer power supply units in the system, and of course more likelihood that these can be fully equipped with standby facilities to protect against service interruptions.

While the industry concentrates on efforts to improve quality and reliability of the cable service there is still a steady demand for the ability to carry more and more channels. This works against these efforts since the greater the channel loading of a system the more the requirements for improved amplifier performance, both in terms of bandwidth capability and distortion levels. It is therefore expedient to examine the reasons for this demand.

Satellite delivery of television programs intended for distribution by cable systems commenced in 1975 with the Home Box Office pay-TV service, and this opened a new era in cable TV and in television itself. No longer was the number of channels carried by a cable system determined by the number of broadcast stations which could be received at the headend, since the satellites used for television distribution were positioned at locations where they could be "seen," and therefore accessed, literally anywhere in continental North America. This effectively eliminated the concept of "local" or "distant" stations which had been the guiding regulatory principle for the previous twenty-five years.

In the two decades which have elapsed since that first venture into simultaneous continental program distribution, the number of satellite transponders available for television transmission has multiplied, and with it the number of television services intended specifically for cable TV reception and distribution. Today there are nearly one hundred of these television services being carried on ten American satellites, and more than forty others on two satellites owned and operated in Canada. At least a dozen of these are pay-TV services such as the original HBO, but the majority are specialty services such as weather, all-news, sports, home shopping, classic movies, etc. In addition to these primarily entertainment and general information services, satellite transponders are also used to distribute nationally in Canada the proceedings of the House of Commons, and in the United States the proceedings of Congress, together with other special interest services, both video and text, plus a number of audio-only radio services.

This multiplicity of services has established a "narrowcasting" environment in which special interest audiences can be targeted simultaneously, with the choice of program to be viewed at any time left to the viewer choosing from a variety of channels without imposing that choice on others. This replaces the previous atmosphere of "broadcasting" in which programming on the limited number of channels had to be designed for a very broad and varied audience, and any choice by the viewer was limited to the number of broadcast channels simultaneously available. It is this ability to cater to the varied interests of many different audiences at the same time which fuels the demand for more and more channels.

The result of this recent and rapid development is that the average cable system serving a metropolitan area or a large city today will have the capacity to carry at least a hundred channels, and is likely to be delivering at least sixty. The majority of these channels will be carrying services received by satellite rather than broadcast services received off-air as in pre-satellite days. Nevertheless, this capacity will not be sufficient for the services expected to be available in the near future and this demand for even more channel capacity is being pressed further by the growing interest in pay-per-view as an alternative to the present forms of pay-TV.

High definition television (HDTV) is another technological development which will further increase the demand for both channel capacity and improved quality in cable transmission. HDTV is the generic term for a system which substantially improves the quality of pictures reproduced on TV receivers over that produced by the current standards used for TV production in North America since the commencement of television broadcasting. The improvement is such as to bring this quality much closer to that of 35 mm film.

The quality of a TV picture can be characterized in two ways — by the number of pixels (horizontal picture elements) contained in a single displayed line, and by the number of lines used to write the picture on the screen. One frame of the picture displayed by the North American colour standard contains about 250,000 pixels in 485 active lines. These are the lines which contain the picture information and do not include the forty lines which comprise the vertical blanking interval during which pulses are transmitted to synchronize the receivers with the transmitter. By comparison a single HDTV picture frame will require about 1.2 million pixels and 1,000 active lines. The key to more pixels and more lines is more bandwidth, and an HDTV system using analog modulation is likely to need at least twice the 6 MHz bandwidth of the present standard television channel.

However, we are now into another technological revolution which will be even more dramatic and far-reaching than that which resulted from the introduction of transistors and solid-state technology into cable TV thirty years ago. This revolution is based on three separate but inter-related developments — fibre optics, digital video, and video compression. Between them they are dramatically widening the boundaries of what is possible on a cable system and completely changing the parameters on which all system design has been based up to now.

Fibre optics is the generic term used for the transmission of data electronically over hair-thin glass fibres using laser-generated light. A laser (Light Amplification by Stimulated Emission of Radiation) has the characteristic of producing coherent or monochromatic light, or light which, unlike other more familiar forms, has a single discrete frequency. So-called "white light," such as sunlight, is a mix of all the colours of the rainbow, each with its own discrete frequency, and therefore covers a wide spectrum of frequencies. Laser light, however, comprises a single frequency and as such can be modulated by a lower-frequency electrical signal much like the modulation of a radio or television carrier. Such a modulated light source can then be launched into an optical fibre and transmitted to a distant receiver where it is demodulated back to a duplicate of the original signal.

A fibre optic transmission path has several major advantages over the more conventional transmission by other media, such as coaxial cable. The most important of these is extremely low attenuation. While a coaxial cable is typically limited at conventional television frequencies to a distance of the order of 2,000 feet (0.61 km) before amplification becomes necessary, an optical fibre can carry a signal for nonamplified distances of as much as twenty miles (32 km). Furthermore, the effective bandwidth of the fibre is very high, extending to many gigahertz, so that the transmission medium itself is unlikely to be a factor in future bandwidth expansions.

The quality of the signal received over a transmission link in terms of noise and distortion generated within the link and the reliability of the service delivered are both related directly to the number of amplifiers through which the signal passes. The very low attenuation of optical fibre, by dramatically reducing the number of amplifiers needed in a given distance, provides a major improvement in reliability while increasing the bandwidth capability and hence the channel-carrying capacity of the link.

Fibre technology was first adopted on a commercial basis by telephone companies during the late 1970s and early '80s. Throughout North America fibre optic cables have been rapidly replacing copper-pair cables, coaxial cables, and microwave systems, both in the long-distance network and for local exchange trunks within cities. In Canada the last section of a 7,000 kilometre coast-to-coast fibre optic cable, stretching from Halifax to Vancouver, was completed in 1990. Such a cable has many advantages over microwave or paired cables for telecommunications. Each hair-thin glass fibre can carry more than 8,000 high quality voice circuits or several television channels, while a typical fibre cable is less than half an inch in diameter, and several of these will occupy less duct space than a typical multi-pair telephone cable and have vastly increased capacity.

The application of fibre optics to cable television was somewhat slower for several reasons. The early applications of fibre optics were designed for frequency modulation in digital mode because this is more easily applicable to voice and data transmission, the primary concern of the telephone administrations. Television transmission, however, both by broadcasting and cable, uses amplitude modulation in analog mode, and the many millions of receivers in public use are designed for this system. The early use of fibre technology in cable TV was therefore largely restricted to the replacement of coaxial supertrunks and multi-channel microwave links between remote headends and distribution hubs from which the trunk system radiates into the service areas. This was due to the high cost and limited availability of both the fibre and the associated opto-electronic equipment, together with the cost involved in the analog/digital channel conversion equipment required at each terminal of the link, which could only be justified on point-to-point links generally at least several miles long.

In 1988 the development of the distributed feedback laser made feasible direct modulation of the light source by broadband amplitude-modulated signals, eliminating the signal processing components necessitated by frequency-modulated links. This substantially reduced costs while permitting the use of fibre cables in the the trunk system itself. In a typical cable system a trunk and distribution feed could comprise at least twenty trunk amplifiers, a bridger amplifier, and three line extender amplifiers to the last subscriber, all in cascade. A single unamplified fibre replacing the trunk cable between the distribution hub and the bridger would eliminate all twenty trunk amplifiers, reducing the total amplifier count in the cascade by more than 80 percent. Since the majority of the noise generated in the transmission path is generated in the cascaded trunk amplifiers, this will dramatically improve the quality of the signal delivered to the furthest subscriber. At the same time this reduction in the number of amplifiers will reduce the probability of service failures, thus improving the system reliability.

These developments boosted the introduction of fibre optics into the cable TV infrastructure at a pace which is now almost rivalling the rate at which the technology was introduced into the telephone network during the previous decade. However, as in the telephone network, where fibre is now used predominantly for the long-distance and inter-office links while the local loops (or "the last mile to the subscriber") are still provided by copper pairs, the application of fibre optics to cable TV is still almost exclusively to the trunk system from the headend to the bridgers, while the distribution feeders and subscribers' drops are still in coaxial cable.

There are valid reasons for both these limitations. In the telephone network, with the necessity for a separate dedicated pair to each and every subscriber, there is a huge capital investment in the local loops. The cost of replacing this investment with fibre would be difficult to justify, at least in the foreseeable future, even though the very limited bandwidth of these loops seriously inhibits their use for more sophisticated services such as multi-channel video and high-speed data. In cable TV there is little doubt that in many systems fibre will eventually replace the coaxial cable and the amplifiers in the distribution system from the bridgers to the subscribers' taps, since this will further improve performance and reliability. However, it is much less likely that fibre will be extended into the individual homes replacing the coaxial drops, especially as, unlike the telephone local loop, these are already wideband and not therefore a limiting factor in the provision of more channels and added services.

The coaxial cable does not itself impose a significant bandwidth limitation, and since the drop is generally not more than one or two hundred feet in length and does not include any amplifiers, it is quite capable, within broad limits, of carrying any number of channels delivered to it by the distribution system. Furthermore, in a typical cable system some 75 percent of the total cable is in the subscribers' drops, the remaining 25 percent being in the trunks and feeders, the portion of the plant which can be most beneficially replaced by fibre. A substantial majority of the homes in North America are now wired for cable TV, and in both the television and the telephone networks the drops into the homes, being located in private property, are the most inaccessible part of the plant for replacement. This represents a major advantage to the cable TV industry in the delivery of future services requiring wideband distribution.

During the last decade, while the telephone administrations have been busy replacing much of their trunk and toll networks by fibre to improve and expand their voice and data services, they have been increasingly interested in the possibilities offered by this upgrading to carry additional wideband services such as video. This has been of concern to the cable TV industry which could foresee serious competition by a financially stronger competitor. However, the realities inherent in the "last mile" to the subscriber now appear to be tending towards a future in which cooperation between the two industries seems more probable than out-and-out competition, with the coaxial drop being used as the entry point for all services into the home even though the trunk and possibly the feeder distribution networks remain separate. This tendency is already evident in the current interest in "multimedia" services in which video, voice, and high-speed data are all carried on a single network.

The second and third factors in the current technological revolution, digital video and video compression, are interrelated and at the same time closely associated with the practical introduction of HDTV. Several proposals have been evaluated by the U.S. Federal Communications Commission (FCC) for an HDTV system to be selected as the standard for the United States and, by implication, for Canada. This process started in 1987 when the FCC invited proposals for High Definition Television and a possible system that would serve as the nation's new technical standard for broadcasting in place of the NTSC system which has been the standard since 1941.

The development of these proposals has been both competitive and rapid, commencing with several systems involving enhancement of the present NTSC analog standard, including a system developed in Japan which had already become almost a de facto standard in that country in the hopes that it may be adopted world-wide. However, the rapid developments since 1987 in digital video technology have resulted in proposals to the FCC involving first several hybrid analog/digital systems, and now the probability that the chosen system will be fully digital. This may set the stage for eventual complete replacement of analog television by digital early in the next century. This will be the biggest change in television technology since the introduction of colour.

The present television world is an analog one. The television picture originates in a video camera which is analog in nature: its electrical output is proportional to the intensity of the light impacting the camera screen being scanned. At the other end of the chain is the receiver, and this also is analog, the brightness of the picture formed on the screen being proportional to the level of the demodulated video signal. Similarly, the accompanying sound originates in a microphone and is reproduced by a loudspeaker, both of which are analog in form.

Thus, the transmission from studio to receiver, whether by broadcast or by cable, has always been, and still is, in analog mode. It is possible to convert the video and sound signals produced in the studio directly into digital mode; in fact many studios are now doing this because in this form they are much easier to switch and to manipulate for commercial insertions and the addition of special effects. They are converted back to analog form for broadcast, and it is the necessity for the signals to be in this form for presentation to today's receivers which is the limiting factor in improving the quality of the pictures through the entire distribution medium.

Since noise and the various forms of signal distortion and electrical interference are also analog in form, they make their presence felt in a transmission path by changing the brightness of the picture on the screen, thus making these interfering effects visible with the picture. Furthermore, these effects are cumulative, and can only cause progressive deterioration as the system lengthens and is subject to further incursions. The practical length of a transmission system using the analog mode is therefore limited by this build-up of noise and distortion, and cannot exceed the point at which they degrade the picture to an unacceptable level.

In contrast a digital signal consists of a sequence of pulses or binary " bits," each of which has one of two states, the presence of a pulse, commonly referred to as a "one," or the absence of a pulse, referred to as a "zero." A group of bits, known as a "byte," defines a numerical code which represents the level or intensity of the original signal at the moment of sampling, and the complete analog signal is represented by a sequence of bytes.

The intensity of the bits in the digital signal is relatively immaterial. All that matters is the ability to detect the presence or absence of a bit and so determine whether it is a "one" or a "zero." Thus, if the intensity of the received signal is modified by noise or interference in the transmission path it does not matter so long as the receiver can determine whether there is a bit there. When the path loss is such that further transmission would render the signal undetectable, it can be regenerated and sent on as a completely new signal with the information it represents intact. By repeating this process the transmission path can be extended in length indefinitely without degrading the resulting reception.

The successful application of digital technology to video depends on the use of video compression. An HDTV video signal with its accompanying sound signal, when digitally encoded, would require a bit rate of approximately 1.5 gigabits per second, and the frequency bandwidth required for its transmission would far exceed the capability of a standard 6 MHz television channel. This signal must be compressed by a factor of about a hundred to allow transmission within the 6 MHz bandwidth, and this can be achieved in several ways.

The method used to convert the signal from analog to digital form is such as to squeeze the maximum information into each bit, but most of the compression is obtained by eliminating the very substantial redundancy that exists in the picture content from frame to frame. Except where there is rapid movement in a scene the video does not change substantially between successive frames. Indeed, if there is no movement over a period of several seconds then several hundred successive frames will be identical, and transmitting every one in its entirety provides no further useful information to the receiver. By detecting and transmitting only the changes from frame to frame the total amount of information to be transmitted can be enormously reduced.

The basic principles of digital video compression (DVC) can also be applied to signal transmission over any medium such as a satellite transponder or cable. In this case the video content of the signal cannot be compressed since this can only be done at the source where the frame to frame changes can de detected, but higher efficiency coding techniques used in the analog to digital conversion can be applied to the modulated carriers such that as many as six or eight signals, each occupying a 6 MHz bandwidth in uncompressed form, can be squeezed into one 6 MHz channel.

The earliest application of digitally compressed television was to transmissions by satellite for direct-to-home broadcasting as described in Chapter 8. The compression of several 6 MHz video channels onto a single satellite transponder enormously increases the capacity of an existing satellite, and goes a long way to compensating for the limited number of available locations for further satellites in the geosynchronous arc over a target area.

In 1995 a group of major Canadian cable systems formed CableSat, a consortium to spearhead the introduction of digital compression technology to cable subscribers. CableSat was intended to act as a provider of digitally compressed satellite-delivered signals to cable companies across Canada, initially in eastern Canada but extending later to the western provinces, offering pay-per-view programming previously available only in major urban areas. In conjunction with Viewer's Choice, a DTH programming undertaking licenced for eastern Canada, Cablesat delivered sixteen channels of pay-per-view movies and specialty services, and divided its development plan into two phases.

In Phase One programming was transmitted by satellite to cable headends in digital format using an eight to one compression ratio, so that only two transponders on the satellite were occupied. At each headend the signals would be converted to analog format and distributed with other channels in the normal way. This method of distribution was intended to provide cable operators of all sizes with a fast response to competitive DTH services.

Phase Two involved upgrading the compression equipment at the uplink to meet new (MPEG-2) standards, and cable operators would then provide digital set-top boxes to their subscribers to enable the services to be offered in both analog and digital versions. In the analog mode the number of services delivered to the subscribers would depend on the spare channel capacity available on each system, and fewer than sixteen channels could be supplied. Once conversion of a cable system to digital mode was completed only two cable channels would be required to carry the full complement of sixteen TV channels delivered from the satellite.

The general application of digital technology to cable TV is most likely to be determined by the introduction of a high definition TV broadcasting service, but the controlling factor will be the subscriber's television set. While the video signals originating at the studio can be readily digitized, delivering those signals to the receiver is an entirely different matter since inputting the signal to the receiver requires a conversion back from digital mode to analog, or the use of an altogether new digital receiver. The receiver is a consumer item and there are millions of them in service with effective lives of, typically, ten to fifteen years.

In this respect the situation is once again similar to that which occurred when colour broadcasting was first introduced. At that time there were several million monochrome receivers in service and it was necessary to introduce the new technology in such a way that these could still receive the new signals and, ignoring the colour content, display them in black and white. The NTSC (National Television System Committee) standards, which were adopted in 1941 to define the television system in North America, were later amended to include colour in such a way that the required compatibility between colour receivers and monochrome receivers was achieved.

However, this did involve certain compromises in resulting picture quality which have become progressively more apparent in recent years as both the technology and public expectations have progressed. These limitations in the name of compatibility are making it a little easier to consider the possibility of adopting a radically new technology which would not be compatible with existing receivers designed to accept NTSC signals. This consideration is assisted by the fact that when it comes to introducing HDTV an entirely different type of receiver will be required, whether the mode of transmission is digital or analog, and if the method of transmission is radically changed this can be designed into these new sets.

By 1995 it was becoming clear that communications technology in the world at large was steadily going digital. Since the inception of television some sixty years ago the world has operated on various production, distribution, and display standards, but all of them were analog systems, and they all, to greater or lesser degree, suffer the constraints that analog techniques impose on both picture and sound quality. The advent of digital technologies was having a profound effect on the way we communicate and entertain ourselves — from digitally-recorded compact discs to the Internet — and telecommunications was already making use of digits in the creation and transfer of everything from consumer credit information to radio and television programs. However, our TV sets, the critical final stop in the broadcasting chain, still use technology that had its origins in the early days of black and white pictures.

With this in mind, in October 1995 the Canadian government created a task force mandated to advise the government on the policy framework required for the transition to digital television and to coordinate this transition in Canada. The task force, which included representatives of Canada's broadcasters, program producers, television service distributors, electronic products manufacturers, and regulators, tackled the various aspects in several working groups. A report was issued in October 1997 under the title *Canadian Television in the Digital Era*.

In its preamble the report said, "while Canada has developed a major television broadcasting and production industry, and has developed what is probably the world's most sophisticated distribution system, the basic quality of the television pictures we see in our homes has not really improved all that much. Certainly colour television is more appealing than black and white, and cable has brought most of us not just more, but also better, pictures, but this is only relative to what we started with — a TV picture that looks all right in a small format, but cannot stand up to today's large screen sizes and to the range of the human eye. Digitization permits literally revolutionary quality improvements in the production, distribution, and exhibition of television pictures and sound, and it allows those signals to be moved about in a fraction of the electronic space now used by analog signals, and free of the many forms of interference to which the analog signals are subject.

"Furthermore, digitization is not something projected to happen some time in the future. Around the world broadcasters are gearing up to introduce fully digital television and, closer to home, the Americans are already preparing to convert their entire television sytem to digital within the next ten years. It was projected that by the new millennium more than half the viewers in the United States will be able to receive off-air digital television services. In fact the American plan called for the introduction of digital TV in the top ten markets by the end of 1998, with smaller centres to follow shortly after. Thus, because of their proximity to major U.S. cities, Vancouver, Toronto, and Windsor will certainly be able to receive U.S. digital television off-air very early on. Some 80 percent of Canadians live in areas where at least one U.S. over-the-air analog signal can be received, and the vast majority of them would shortly have the option of receiving American digital TV signals from the same sources."

The task force concluded that Canada's interests would best be served by seizing the initiative and developing a comprehensive, coordinated digital transition plan to ensure that our broadcasting system remains strong and vibrant, providing a full range of competitive services designed to meet Canadian needs, yet flexible enough to meet the uncertainties involved in predicting the future. For this purpose its report proposed the following steps:

> 1. Canada should formally adopt the ATSC Digital Television Standard (A/53) for terrestrial transmission as defined by the Advanced Television Standards Committee of the U.S.A. and approved by the FCC.

The basic purpose of the A/53 standard is to provide a common technological environment within which the various digital TV formats and additional data services can coexist.

2. All over-the-air broadcasters should be granted a digital licence and should be required to implement it by the end of 2004. Broadcasting transmitters serving the largest markets (e.g. Montreal, Toronto, and Vancouver) should begin digital transmission by the end of 1999, followed as soon as possible by stations in the next largest markets (e.g. Edmonton, Calgary, and Ottawa) with the objective of being digital over the air in all markets by the end of 2004.

3. When a broadcaster or other program provider makes available a digital TV signal within the new standard the superior quality and format of that signal should be passed through to the consumer by all Broadcast Distribution Undertakings, and all BDUs should be fully digital-capable by the end of 2004.

4. All analog over-the-air transmission should cease at the end of 2007, but, because of the uncertainties of the program timing, the feasibility of this target should be assessed annually beginning in 2004.

5. Digital television services must be clearly superior to existing analog transmission and, at a minimum, all digital pictures should be in the 16 to 9 (wide screen) aspect ratio with a resolution at least equivalent to the existing 525-line analog standard.

6. By the end of 2007 two-thirds of each broadcaster's schedule and two-thirds of new Canadian content productions should be available in the HDTV format.

Subsequent review of these proposals indicated that the suggested time scale for their adoption might be unduly optimistic. However, this broad policy provided the mechanism for a controlled transition from a totally analog system to a totally digital system, while still leaving many problems of detail in its implementation to be resolved.

One of the most important of these was the need to adapt the TV sets to the new system. Sets will shortly be on the market suitable for digital reception. Design and production of these had been delayed because the standards to be adopted for a high definition system had not yet been finalized. During the transition period provision must be made for continuity of service to the existing sets, and this compatibility can be achieved in two ways. Digital signals can be converted back to analog form at each TV set using a set-top box similar to those presently used to extend the channel capacity of the sets but including digital-to-analog conversion in addition to channel tuning.

Alternatively, the conversion can take place at some intermediate point such as a cable TV headend, or the program material can be duplicated by the broadcaster for the duration of the transition period using a second channel. It appears likely that broadcasters will be allowed, on a temporary basis during the transition period, to broadcast in duplicate on two channels, one in analog and the other in digital format, but the extent to which this is practical depends on the availability of sufficient broadcast channels for this purpose at each transmitter.

Duplication of program material in both formats is more easily achieved on a cable system, since the signals received in digital format at the headend can be converted at that location and then transmitted over the system on a second channel. In this way all program selections would be available to the subscriber in either digital or analog format, and the subscriber would select the channels according to the requirements of his or her set. This solution also requires the availability on the cable system of spare channels, but with the use of digital video compression, capable of increasing the number of channels by a factor of six to eight, this is less of a problem than the limited number of over-the-air broadcast channels.

It was stated earlier that a high definition TV composite signal when digitally encoded would require a bit rate of approximately 1.5 gigabits per second and that, in order to accommodate this in a standard 6 MHz TV channel, it would have to be compressed by a factor of about a hundred. Much of this compression is accomplished by removing the inherent redundancies in the original picture, but part of it is obtained by the method used to encode the analog signal into digital form. Typically this encoding can achieve a reduction by a factor of six to eight, and this type of compression can be applied to the analog signal at any point in the transmission chain at which the signal is being converted from analog to digital format.

If a number of standard NTSC signals in the usual uncompressed form as broadcast are digitized using this degree of digitizing efficiency, it is possible to accommodate them in a single 6 MHz-wide channel. If applied to all the channels carried on a cable system this would increase the capacity of the system by this factor. Thus, a typical system designed to have say a thirty-channel capability carrying standard NTSC signals could increase the number of channels carried to some two hundred or more by digitizing and compressing all the channels without any need for increased bandwidth. This capability of substantially increasing the channel-carrying capacity of the coaxial plant without actually increasing its bandwidth will complement the much wider bandwiths available on the fibre-optic portions of the future systems, and will enable cable systems designed for present demands to expand the services offered even further without the need for drastic rebuilding.

The question is often asked, "Why would any viewer want two hundred or more channels? Surely this is taking choice to the extreme." The answer to this question lies in the changes taking place in the nature of television itself. During the first decades of this public service it was essentially "broadcasting" in the broadest sense of the word. Because there was a limited number of program producers and a limited number of channels on which the material could be transmitted to the viewing public, the programs had to be conceived with the widest appeal. Today the number of channels available, both by satellite and on terrestrial systems, has multiplied, and this has allowed the addition of many specialty services catering to much narrower audiences — in fact "narrowcasting" rather than "broadcasting."

The availability of many more channels on cable puts the choice in the hands of the viewer rather than the program originator. An appropriate comparison might be a bookstore offering a wide choice of magazines. Any one customer may be interested only in the subjects covered by one or two of the magazines, but the wide choice displayed caters to all tastes.

Choice is not only a matter of different programs, but also time of viewing. Offering a program at a specific time, as with the present method of programming, is no choice to the individual viewer if the time at which it is offered is inconvenient. Repeating the program at other times is a partial remedy but still may not be convenient, and inevitably reduces the choice available to others who have already watched the original showing. However, if a multitude of distribution channels is available, then that program can be repeated several times without inhibiting the choice available to others.

Applied to pay-TV the availability of a group of, say, eight channels for one program offering would enable a program in that group, for example a two-hour movie, to be run simultaneously on all eight channels with starting times staggered every fifteen minutes. Viewers would thus have the ability to watch the program at times convenient to them and never have to wait more than fifteen minutes for the start. If several such groups of channels were dedicated to this type of programming then a number of movies could be run simultaneously, providing not only a choice of program material but a choice of viewing times as well, and this would come close to the ideal of video-on-demand. Since compression factors of the order of eight to one are already possible it will be seen that the availability of even ten additional standard 6 MHz channels with video compression added to a cable system would greatly widen the scope of a pay-TV offering in terms of customer convenience.

This type of programming has already been field-tested on a limited scale by the Canadian pay-TV channel "First Choice — The Movie Network" on a test system in Ontario. Using video compression, movies were offered at a variety of start times on a group of four channels, and the commercial results were sufficiently encouraging that it is expected that multiplexing in this form may become a regular feature of The Movie Network once the technical facilities are in place on a sufficient scale.

There is no doubt that a further revolution is under way now in cable television. During the first quarter century of cable TV's development, from 1950 to 1975, the industry was technology driven, technical changes paving the way for widened programming activities which never quite met the promises held out by the technical developments. This situation changed drastically when Home Box Office commenced national distribution of pay-TV by satellite in 1975. This was followed by a rapid proliferation of satellite-distributed program material intended for ultimate distribution by cable. The era of more channels than program sources quickly reversed to the point where, even with further technical evolution in the capability of cable systems, the introduction of such multiplexing ideas as those just discussed was again dependent on a further substantial increase in channel capability. This renewed pressure for still more channels, added to the demand for higher definition and improved quality, has led to the latest technological developments of fibre optics, digital transmission, and video compression, and these are in the process of changing completely the entire technological and sociological basis of the cable TV industry, and indeed of television itself.

Appendix

Growth of Cable TV 1974–1998

Chapter 8 describes the growth of cable TV in Canada from the early 1980s after the introduction of colour broadcasting and the start of program distribution on a continental scale by satellites, but this growth is most effectively illustrated by the statistics covering the period from 1974 to the present.

Year	Homes passed	Total Subscribers	% Penetration
1974	4,045,000	2,561,000	63.3
1976	4,706,000	3,143,000	66.8
1978	5,536,000	3,776,000	68.2
1980	6,111,000	4,339,000	71
1982	6,600,000	4,933,600	74.7
1984	7,100,000	5,389,000	75.9
1986	7,700,000	6,004,700	78
1988	8,340,000	6,624,200	79.4
1990	8,948,000	7,163,300	80
1992	9,555,900	7,691,300	80.5
1994	9,936,700	7,994,800	80.6
1996	10,273,300	8,206,819	79.9
1998	10,336,817	8,349,589	80.8

Cable TV Subscribers by Category — 1998

Residential		
Homes	6,828,481	81.7% of total subscribers
Apartments	988,554	11.9%
Total residential	7,817,035	93.6%
Commercial		
Hotels/Motels	282,999	3.4%
Health facilities	110,691	1.3%
Educational facilities	13,077	0.2%
Other commercial	125,787	1.5%
Total Commercial	532,554	6.4%
Total Subscribers	8,349,589	100%

Cable TV Subscribers by Province — 1998

Province	Homes passed	Subscribers	Penetration
Newfoundland	174,720	138,401	79%
P.E.I.	38,760	32,199	83%
Nova Scotia	289,201	235,596	81%
New Brunswick	216,028	188,434	87%
Quebec	2,768,146	1,856,398	67%
Ontario	3,787,698	3,007,015	79%
Manitoba	378,497	274,953	73%
Saskatchewan	291,840	205,772	71%
Alberta	901,809	712,199	79%
British Columbia	1,467,643	1,149,478	78%
Yukon	7,590	5,833	77%
N.W.T.	14,885	10,757	72%
Total	10,336,817	7,817,035	76%

Specialty and Pay-TV Services

Since 1984 the CRTC has authorized a total of fifty-four "specialty" and pay-TV program services for carriage on cable TV systems, wireless distribution networks, and direct-to-home satellite distribution services in Canada. These specialty services cater to the interests of audiences with more specialized interests than the audience to which the broadcast networks are directed, and include pay-television and pay-per-view services.

Specialty TV services:
English language 27
French language 9
Third language 5
Services with French/English feeds 2

Pay TV services:
English 5
French 1

Pay-per-view services:
English 4
French 1

The Commission's incentive in authorizing these specialty services has been to foster the development of original programming and ensure a stronger presence of Canadian content in both English- and French-language markets in the face of increasing globalization of communications. They also respond to consumer demand for new, affordable, and high-quality TV options. Although fifty-four have been authorized to date it was recognized that not all of them could be activated immediately, particularly on cable distribution systems, because of the limited channel capacity dictated by the existing analog technology. However, with the deployment of digital video compression the resulting increase in channel capacity will enable distributors to offer their subscribers a much greater number of services. Some of these may be included with the broadcast services as part of the "basic service" offering, but the majority will be offered as "discretionary services" for an additional fee over and above the price of basic service.

Following is a list of the fifty-four specialty and pay-TV services authorized from 1984 to 1996.

Date	Service	Language
1984	Fairchild TV (formerly Chinavision)	English
	MuchMusic	English
	Telelatino	Ethnic
	The Sports Network (TSN)	English
1987	Canal Famille	French
	Weather Network/Meteomedia	English/French
	Musique Plus	French
	Newsworld	English
	Le Reseau des Sports (RDS)	French
	TVS	French
	Vision TV	English
	YTV Canada	English
1992	Talentvision (formerly Cathay)	Ethnic
1994	Bravo	English
	Canal D	French
	The Discovery Channel	English
	The Life Channel	English
	New Country Network	English
	Le Reeau de l'Information (RDI)	French
	Showcase	English
	Women's Television Network	English
1996	Canadian Learning Channel	English
	Le Canal Nouvelles	French
	Le Canal Vie	French
	The Comedy Channel	English

CTVN	English
The History & Entertainment Channel	English
Home and Garden Television	English
MuchMoreMusic	English
Musimax	French
Odyssey	Ethnic
Outdoor Life	English
Prime TV	English
Pulse 24	English
Report on Business (ROB-TV)	English
S Regional Sports Service (S)	English
South Asian Television (SATV)	Ethnic
Space: The Imagination Station	English
Sportscope Plus	English
Star-TV	English
Talk-TV	English
Teletoon	French/English
TreeHouse TV	English

Pay-TV services currently available

The Family Channel, English
MovieMax, English
The Movie Network, English
MoviePix, English
Superchannel, English
Super Ecran, French

Pay-per-view services authorized for DTH and terrestrial distribution

Canal Indigo, French
Power DirecTicket, English
Home Theatre, English
Sports/Specials, Pay-Per-View, English
Viewer's Choice, English

THE AUTHOR

Ken Easton spent the first fourteen years of his professional life from 1933 to 1947, including the war years, with the British Post Office in telephone engineeering. In 1947 he joined Rediffusion Ltd., a leading British radio relay company, as Engineer-in-Charge of the London operation, and was personally involved in the development of the first television relay and community antenna television (CATV) systems. Between 1950 and 1953 he organized and headed a new Television Department at Rediffusion's head office in London, and acted as a consultant to the regional companies in the application of television to many of the radio relay systems then being operated by Rediffusion throughout England and Wales.

In 1953 he transferred to Canada as chief engineer of the major cable television system then under construction by Rediffusion in Montreal. In 1960 he joined Famous Players Canadian Corp. in Toronto to take charge of the engineering of the experimental pay-TV cable system being installed there by International Telemeter, a subsidiary of Paramount Pictures. In 1963 he was appointed Vice President, Communications, by Famous Players, and was very active during the next seven years in developing major cable TV systems in many parts of Canada.

In 1969 a federal government Order-in-Council precluded Famous Players from significant ownership participation in any broadcasting and cable TV licences on account of its partial foreign ownership. Ken Easton then resigned from the company and commenced private practice as a cable TV consultant. In this capacity he was active in developing major cable systems throughout Canada and the United States, including many using microwaves and later satellites, until 1990 when he finally retired.

Ken Easton was closely involved as a founding member in the establishment of the Canadian Cable Television Association in 1957 and, as part-time Secretary, in its operation for the first ten years until 1968. In this capacity he was personally involved in the many discussions with representatives of the government leading up to the Broadcasting Act of 1968 and the establishment of the Canadian Radio-Television Commission (CRTC) as the regulatory authority over cable TV as part of the broadcasting industry.

Ken Easton is a member of the Institute of Electrical Engineers (IEE) of the U.K., a life member of the Institute of Electrical and Electronic Engineers (IEEE) of the U.S.A., a fellow of the Society for Cable Television Engineering (SCTE) of the U.K., and is registered in Ontario, Canada, as a Professional Engineer (P.Eng.).

The author and publisher acknowledge the cooperation and support of the Canadian Cable Television Association.